My Search f
in the year 2003

To Pat , Peter Knight.

Very best wishes,

John Caussey 16 April 04

BOOKS AVAILABLE from:-
Rose Cottage,
34 Overgang,
Brixham, Devon,
England, TQ5 8AR
Telephone (01803) 854554
PRICE £9.95

ERRATA :
p.94 - 3 lines up -
"tree" should read "free"
p.66 - line 19
"Page one" should read "Page 42"
p142 Section 528
"part 3 should read - suffering ends with
overcoming desire - NOT Suffering ends
with desire to overcome"

My Search for the Truth in the year 2003

by John Cawsey

Trafford Publishing Ltd

Design and typesetting by Mike Harrington
MATS, Southend-on-Sea, Essex

Printed in Victoria, Canada

Note for Librarians: a cataloguing record for this book that includes Dewey Classification and US Library of Congress numbers is available from the National Library of Canada. The complete cataloguing record can be obtained from the National Library's online database at:
www.nlc-bnc.ca/amicus/index-e.html
ISBN 1-4120-1640-1

TRAFFORD

This book was published on-demand in cooperation with Trafford Publishing. On-demand publishing is a unique process and service of making a book available for retail sale to the public taking advantage of on-demand manufacturing and Internet marketing. On-demand publishing includes promotions, retail sales, manufacturing, order fulfilment, accounting and collecting royalties on behalf of the author.

Suite 6E, 2333 Government St., Victoria, B.C. V8T 4P4, CANADA
Phone 250-383-6864   Toll-free 1-888-232-4444 (Canada & US)
Fax 250-383-6804   E-mail sales@trafford.com   Web site www.trafford.com
TRAFFORD PUBLISHING IS A DIVISION OF TRAFFORD HOLDINGS LTD.
Trafford Catalogue #03-2017   www.trafford.com/robots/03-2017.html

10   9   8   7   6   5   4   3

Chapter Headings

1   Understanding some astronomy, the universe, cosmology.

2   The formation of our solar system, the earth, the moon, mapping the earth, the Magnetic core, the magneto sheath, volcanoes, tectonic plates, the earth's crust, times life appeared on earth, composition of the earth and its atmosphere, the chemistry and composition of matter, the elements, photosynthesis, the oxygen, sulphur, carbon and nitrogen cycles, the elements, Isaac Newton three laws of motion, gravity physics, Einstein's theory of relativity, the electromagnetic spectrum.

3   Meteorology, how the weather is formed, meteorological definitions, interpreting a weather map, global warming, global cooling, El Nino, hurricanes, atmospheric pollution.

4   The discoveries of early man from 40,000 BC. Where did his beliefs come from? Can we learn anything from our ancient ancestors?

5   The human brain, comparison to animal brains, memory, learning, consciousness, comparison to a computer, the brain's limitations.

6   Unexplained phenomena, ghosts, poltergeists, faith healing, mediums, spirits, extra sensory perception, unidentified flying objects, aliens.

7   The progress of scientific discoveries, a list of inventions from BC time to 2002 'completion of the human genetic code.' A few conclusions are reached.

8   Exploring common ground for uniting different religions, 20 discussed, early religious rituals and beliefs, the existence of god, the formation of a world church of god with a good moral code.

9   The future for planet earth – catastrophes discussed under two headings – 1) Dangers from the sky, 2) Dangers within our atmosphere.

10 The environment a) Solving man's problems in developing and adapting to it – Ecology. b) The effect new discoveries will have on mankind in the next 50 years.

11 Further thoughts and conclusions.

About the author

John Cawsey has seafaring blood from generations of sea captains, from his grandfather, great grandfather, and generations he has traced back to two pilgrim ships in 1605 and 1609, which sailed for Virginia to colonize America. The story of the seafarers from Appledore and Bideford, trading under sail, are well documented in the National Maritime Museum, Greenwich, London.

Going to sea on his father's yacht at an early age in the Bristol Channel, it wasn't long before John captained a sailing vessel for a month's cruise in the Bay of Biscay, and then purchased his first yacht. This has led to a teaching career, and a licence issued by the Royal Yachting Association on behalf of the DTI to John's sailing school, Cutlass Sailing School in 1969. Further night school teaching and the development of course notes for day skipper and yacht master subjects, and John wrote a book of over one thousand pages covering all the subjects necessary to teach day skipper and yachtmaster subjects by distance learning, and continues to teach today as a yachtmaster instructor.

In 1979 John built a 36-foot yacht in his front garden and sailed it across the Atlantic and back. John has since sailed the yacht back to the Caribbean.

As a teenager John showed an interest in astronomy and joined a local astronomical society. He built his own telescope. This was the beginning of his search for the truth. What was out there in the sky?

Passing 'A' Level subjects in physics, chemistry and biology, pointed to the start of a career. At the early age of 5, John learnt to play the piano, and hoped his talent would allow him to embark on a musical career. By the age of 18 his musical career was abandoned, when it was evident that the life of a concert pianist would be too difficult. John entered Bristol University to study medicine, and 6 years later qualified as a dental surgeon. The secrets of life are locked in brain power, the controlling centre of living things. To unravel the secrets of the brain, John studied the brain and practised hypnosis, to try to understand the difference between our brain and a computer, and how we could develop a computer to think for us.

Acknowledgements

Professor Arthur Collie

Many thanks to Arthur for proof-reading the book and advising me.

A retired professor of electrical engineering, an authority on robots, making robots to work in atomic power stations; a great scientist and thinker. A great inspiration and guide to students attempting a university career in engineering.

Susan Beale

A scientist and book editor for 30 years, Susan has much experience and advice to give me.

Introduction

Hi, Chimp! We used to be brothers 4½ million years ago. In all this time our genes have only changed by 1.5%. Do you remember you once said, 'What is the universe about?' Well, today the Year 2003 we have found out a lot about our universe. That's why I am writing a short book for laymen to read. The new discoveries in the last 100 years will change our lives in the future. These discoveries have unlocked secrets that will enable scientists and mathematicians to expand their knowledge at an accelerating rate, not previously known to mankind. If my book stimulates you to think about the future for mankind on earth, perhaps you can help for a better future. Yes chimps, we still like things in common. We want to be loved and share our happiness – that gives us a feeling of stability.

Modern man stems from a common ancestor about 250,000 years ago. It is a wonder we haven't become extinct when the expectancy of life was only 18 years. In America in 1920 it was 49, in the year 2003 it is 76. In 2050 it could be 150, provided we can keep our body cells young and ourselves healthy. Microbiology has discovered how we are made, and the future of medical treatments will be revolutionized. Particle physics has unlocked the secrets of matter, making space travel possible and also world communication (The Internet).

The history of planet Earth is locked up in the ground and rock strata, and under the sea with its ever rising levels. By understanding the catastrophes of the past, perhaps we can plan better for the future. We have only just started to explore our universe.

Come back in 1,000 year's time and we will have many more answers. I don't think we will ever know it all.

Five billion years ago our solar system started to form. From a ball of fire came Planet Earth, cooling to a level where it could contain water and life. Eventually this ball of fire produced life in abundance, the beauty and diversity of creation, the colours, the flowers the insects, the birds, the fishes, the animals, and finally man. The driving force of nature is to create and to change. Nothing is forever. Are all our efforts in vain? No, we must stay ahead in knowledge, in physical and mental fitness and production effort. The alternative is unbearable extinction. We still have a lot to learn, many secrets to unravel. Remember the common denominator from the Universe; from a ball of fire sixteen billion years ago came the infinite beauty and diversity of creation.

Q   **What is happening to the Universe?**
A   It is expanding in every direction.

Q   **When will it stop expanding?**
A   When the energy from the Big Bang is used up.

Q   **How do we know when this happens?**
A   Stars expand until the nuclear reactions are used up and gravity is not supported by the nuclear reactions. The unsupported weight of matter causes the stars to implode into themselves into a tiny ball. All is sucked in. Massive gravitational forces prevent light escaping. It is like a giant Hoover cleaning the universe. These black holes unite and eventually whole galaxies collapse. The universe implodes. The speed of implosion causes matter to become energy. Eventually all will be a ball of energy.

Q   **Where will this ball of energy be?**
A   There is no space. Space is only the distance between objects. There are no objects.

Q   **How long will this state exist?**
A   There is no time. Time is related to mass/energy/speed.

Q   **When did the big bang occur?**
A   16 Billion Years ago.

Q   **Was matter made then?**
A   Yes, slowly. Photons of light took 300,000 years to make.

Q   **Does this mean the big bang was in darkness?**
A   Yes.

Q It must have been an enormous bang?
A Yes, the predicted background radiation from the big bang has been detected.

Q How much matter had been created?
A Only 30% of the possible matter appears present. Einstein's formula $E = MC^2$ says 70% must be energy.

Q Where is the energy?
A The energy is causing the universe to continue expanding. Imagine a balloon being blown up. If all the galaxies were dotted on the balloon every galaxy would be slowly moving away from each other: Those furthest away from us would appear to be moving away very rapidly.

Q Could the big bang have occurred before?
A Yes, but we will never know.

Q What causes this energy to explode into matter?
A Energy and matter are unstable, change is normal.

Q What is change?
A Instability.

Q Does this mean we could have had a different universe in the past?
A Yes, the forces of the big bang could have produced a different set of forces of nature.

Q What are the forces of nature we observe today?
A i Gravitation
ii Electromagnetism
iii Small forces
iv Big forces – e.g.) nuclear reactions and forces holding particles together.
v Action/reaction/no action.

Q How much do we know about the Universe in the year 2003?
A A lot more than we did 100 years ago. We have a good knowledge of particle physics and how life evolves, its very 'make up'.

Q **What do we see and understand when we look at the stars?**
A The number of stars in our galaxy, the number of galaxies, the infinite number of possible planets orbiting their stars, the cloud gases, floating free in space with all the chemical building blocks of life.

Q **What does this mean? Are we alone in the Universe?**
A No, the chances are there are millions of other inhabited planets at different stages of development. Some much more advanced than ours.

Q **Can we expect to be in contact with them?**
A At the rate scientific knowledge is expanding, it will happen in the future.

Q **If they are much more advanced than us would they not have visited us in the past?**
A Quite probably.

Q **What do you think they thought of us?**
A I expect their planet was burning up with their expanding sun, and they are looking for somewhere else to go.

Q **If matter is not being created fast enough, there must be very little matter toward the edge of the expanding universe?**
A Yes, so little matter the forces of gravity are non-existent.

Q **Does this mean the more the universe expands, the less the matter and the less the gravity?**
A Yes, matter will eventually each a state of chaos where there is no gravity a vacuum.

Q **Can the universe totally contract, implode into a ball of energy then?**
A No, not completely. There will always be rarefied matter in a chaotic state outside the forces of gravity.

Q **What keeps them separate?**
A Action and reaction, and no action.

In analysing the evidence I have indicated three lines of approach.

1)   The scientific experiment – that the experiment can be reproduced and demonstrated to produce the same result, the concept that it is proved in scientific terms.

2)   The collective experiences of many people's observing experiences not easily explained in scientific terms, but having repeatedly predictable patterns.

3)   The concept of the judge summing up a case – the weight of evidence is in favour of ' such and such.' This broader approach at times can border on scientific analysis e.g.) I) The weight of evidence is for mathematical laws that govern the universe. 2) The weight of evidence suggests everything in the universe is made from elements in the stars. 3) The weight of evidence suggests our brains could be much larger and we could be very much more intelligent because we only use a small part of the electronomagnetic spectrum. 4) The weight of evidence suggests the building blocks of life which have been identified in cloud gases in the sky, and DNA, have been in existence for billions of years before planet earth was formed. 5) The weight of evidence suggests life is in abundance in the universe.

With the accelerating rate of knowledge, today is an exciting time to be living. John's analytical approach to solving the mysteries of life has produced remarkable conclusions. He has kept the book at a level of knowledge only requiring the basic understanding of science, so that the readership of this book can reach most people.

Precis of the Chapters

CHAPTER ONE

Is educational, explaining the 'Big Bang' Theory, how matter was made, how the stars and galaxies were formed and at approximate times, leading to the formation of our solar system. Everything is made inside a star all the elements. Cloud gases in the sky have been analysed and contain the chemicals, amino acids and building blocks that make life and DNA. If a planet contains water, DNA and life will probably form. Life is in abundance in the universe.

Conclusions

There are laws governing the universe. Who made the Laws, the Mathematics, the Physics, Chemistry and everything else we do not know. Scientists are spending billions of dollars every year on research, space exploration and trying to contact other civilizations.

CHAPTER TWO

Is educational, explaining how our solar system formed, planet earth, and the stages it went through to cool, form a crust, form continents and oceans. Some chemistry and physics are explained.

Conclusions

That the same procedures of planetary cooling occur throughout the universe, where water is present, life forms are likely to be present. There are billions of inhabited planets throughout the universe, some much more advanced than ours, the only common denominator is the beauty and diversity of creation, and the tendency for nature to have a stable form to return to. The evidence suggests a master plan of design. Perhaps this design came from a computer and the design was built into a material form.

CHAPTER THREE

Meteorology is explained. We are in a period of global warming. 10,000 years ago was the end of the last ice age when Europe was frozen and polar bears roamed the South of France. What factors control the temperature of the earth? Is man contributing to the heating of the earth by greenhouse gases? We must safeguard our atmosphere. It is precious. If our atmosphere is badly polluted we will cease to exist.

CHAPTER FOUR

In our search for the truth can we learn anything from our ancestors? Was there a great civilization in the past? The human brain was fully formed 100,000 years ago. Mathematics was understood a long time ago. The knowledge of precession was understood. Temples were aligned to the position of the stars. How did this knowledge become widespread from Egypt to South America? What did ARISTOTLE have to say about creation and the existence of life after death?

Conclusions

We can learn from our ancestors.

CHAPTER FIVE

Why did we need such a complex brain fully formed 50,000 BC and nearly fully formed 150,000 years ago, when life was nomadic? Animals do not need to make great conscious decisions to plan ahead. Our brain has many inherited faults from several million years of development. Why do we have emotions which can be strongly stimulated by going to a music concert? How clever is our brain? The brain is only able to receive a small part of the electromagnetic spectrum, so we can only analyse a very limited amount of input.

Conclusions

We are never going to find out the truth with our brains. We need to build computers to access the knowledge and will they only be able to talk to other computers? Our brains can be accessed through psychokinesis. This is discussed in Chapter 6. Have our brains been altered and for what purpose?

CHAPTER SIX

Explains the current thinking on unexplained phenomena under the heading 'The Collective Experiences of many peoples observing experiences not easily explained in scientific terms, but having repeatedly predictable patterns.'

The reception of infrasound by animals leading to their ability to react to the presence of ghosts and spirits is discussed. Certain people have this ability. John gives accounts of two experiences from deceased persons. Examples are given of poltergeist activity and psychokinesis. The continuing presence of unidentified flying objects is mentioned. Conclusions reached are on the experiences of many people's observations. It is suggested there are forces operating on a level we cannot interpret. Maybe one day we can explain them in scientific terms. There appears to be another level of communication we cannot tune into, and on a different time basis.

CHAPTER SEVEN

A list of inventions from BC times to the present day is not only educational, but illustrates how scientific progress moves forward slowly, stage by stage, one door opens the way to another. Today so many doors have been opened, the rate of acceleration of scientific knowledge is enormous, and there are so many new avenues to go down, in the quest for the truth.

A few conclusions are drawn at the end of this chapter.

1   There are laws governing the universe.
2   The common denominator is the beauty and diversity of creation, and the inherent tendency for chemistry in nature to return to the neutral or stable form.
3   Everything is made in the stars, and cloud gases in the sky have been analysed as containing the chemicals, amino acids the building blocks of life.
4   Where there is water on a planet, life forms will exist, some more advanced than others, and the universe is studded with life. Planet earth is relatively new in the universe.
5   That our brains can be accessed through psychokinesis and may have been altered and modified in the past.
6   There are forces acting on a level we cannot access.
7   That everything may have been made on a computer, and later built from material elements.
8   The way ahead is to build super computers to take over from our limited brains.
9   Scientists are spending vast sums of money on space exploration and trying to contact other civilizations. I believe we will make contact with other civilization in the next one hundred years.

The remainder of the book chapters 8, 9 and 10 are devoted to how to stop planet earth becoming destroyed by war, population explosion and atmospheric pollution. Chapter 10 discusses the effects new discoveries will have on mankind in the next 50 years. Chapter 11 concludes with further thoughts and conclusions.

Chapter 1

Understanding some astronomy –
The Universe

In the beginning there was a very big bang. This happened 16 billion years ago. At very high temperatures and speed, particles were ejected called Protons and Neutrons. Many other particles were made, for example hydrogen, carbon, nitrogen and oxygen atoms. There followed a period when cloud gases condensed and stars were borne, which radiated heat and light from their nuclear reactions. Stars get their energy by fusing hydrogen to helium. Smaller aggregates of cloud gases and rock cooled to form planets in orbits around their stars. Billions of stars can make up a galaxy, and galaxies rotate about a central axis. There are billions of galaxies.

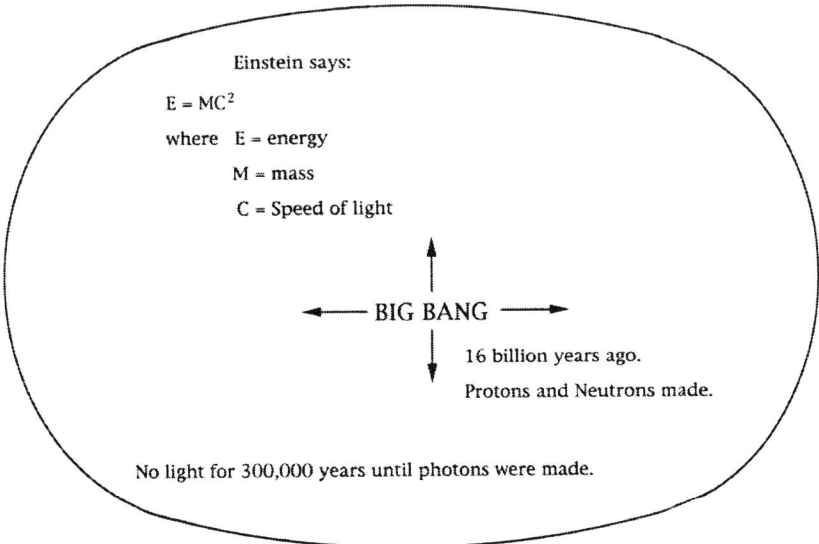

Einstein says:

$E = MC^2$

where E = energy

M = mass

C = Speed of light

←——— BIG BANG ———→

16 billion years ago.

Protons and Neutrons made.

No light for 300,000 years until photons were made.

Like dots on a balloon being blown up we are all expanding away from each other and distant objects appear to be receding away very fast.

Distances across the universe are so great they are measured in light years, i.e. the distance light takes at 186,000 miles per second to travel in one year. Our galaxy is 80,000 light years across and our solar system is about ⅓rd from the outer edge or about 27,000 light years from the galactic centre. The period of rotation of our sun around the galactic centre is 250 million years. The galaxy rotates as a result of the balance between the centrifugal force produced by its rotation and gravitational force. There is a galactic magnetic field. The galaxy is about 8 billion years old. Our solar system is about 5 billion years old. In about 1000 million years time our sun will expand into a red giant and boil away our ocean. We will have to leave and go to a more distant planet that contains water and can support us.

Our galaxy started forming 8 billion years ago

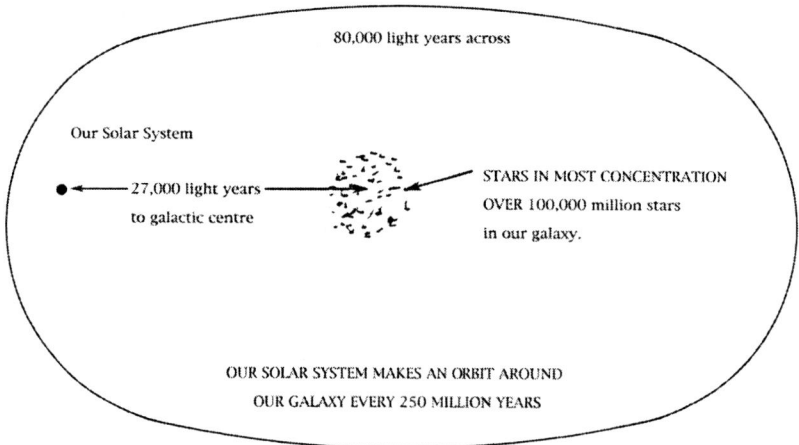

80,000 light years across

Our Solar System

27,000 light years to galactic centre

STARS IN MOST CONCENTRATION OVER 100,000 million stars in our galaxy.

OUR SOLAR SYSTEM MAKES AN ORBIT AROUND OUR GALAXY EVERY 250 MILLION YEARS

The largest stars form at the centre of the galaxy where there is greatest mass. Some of the stars are very large, 1000 times bigger than our sun. An average star lasts 10 billion years. A very large star can burn off its 'energy much faster in only a million years. The weight of its matter cannot be supported by the nuclear reactions, and the star can collapse under the weight of its mass into a neutron star, only a few miles across and very dense. When a star implodes, matter becomes so condensed and the gravitational forces so large that matter near it is sucked in by the immense gravitational forces, including light, hence the name the black hole.

Recent research from many observatories have supported the big bang theory as the beginning of time and events. The amount of matter in the universe can only account for 30%, where has the rest gone, or is 70% energy? The total mass of the universe does not appear to be changing. If black holes swallow up matter, then new stars must be formed continuously. From our earth all galaxies appear to be receding from us. Eventually when the energy of the big bang is used up will the universe stop expanding and start contracting? If there is an edge to the universe, it will be at a very low temperature and rarefied space, with weak or no rotational forces. In such rarefied space there will be no forces of matter attracting matter. Matter will be in a state of chaos. I doubt if in the time since the big bang 16 billion years ago, the outer edge of the universe (if there is one) has ever rotated once. What happens if some galaxies rotate to the right and some to the left? Will they meet at some point and collide?

If expansion is still occurring you would expect less stars to be visible and the night sky getting darker. This does not seem to be the case. if contraction is occurring in every direction the density of matter should be getting more and the night sky getting lighter. There appears to be no change. If expansion stops, contraction will eventually occur if there is enough density of matter. Such a contraction may take some time to get going, but would accelerate as gravitational forces increased. Is a very big black hole going to swallow us up from which we cannot escape? I doubt it. I believe there is a critical size above which a black hole cannot increase, and energy from black holes causes new stars to form.

What caused the big bang to happen is not known. There was no time, no previous event. Matter or energy has always existed. What a difficult concept to understand. What is beyond the edge of the universe?

Space only exists between objects. Where there are no objects there is no space. Sometimes we do not have words to express ourselves. Mathematicians have formula to express their thoughts. To understand what is beyond the universe let us define space. Space is described as an unlimited expanse in which all objects exist and move. A vacuum is defined as an empty space from which all or most air or gas has been removed. Neither of these definitions helps us to understand what is beyond the universe. Let us invent a word. I will call it SPACUUM, and I will define it as a space that cannot be measured by distance and time. For the universe to expand into it we have to create the right condition to pass through this event horizon.

Cosmology is the science of trying to discover the nature and origin of the universe. There are cloud gases in the sky containing all the building blocks

3

of life. If there are so many billions of planets in the universe, is life in abundance in the universe? Scientists seriously believe so, and are trying to detect signals across the universe hoping intelligent life forms are transmitting. They believe contact will be made in the next 100 years. The earth's atmosphere hinders experiments from the ground and a space telescope has been in orbit for some time, collecting valuable data. Satellites are conducting different experiments. Rockets have propelled spacecraft to the moon, planets and towards the sun. Another 20 years time will give scientists the chance to analyse all this data. To find water on a planet in our solar system would be exciting, where other life forms may have existed in the past, or are still surviving.

Scientists sometimes make an exciting new discovery, only to be disappointed later when too many variables come into the equation which are unexplainable. Many years ago when the microscopic nature of matter was first discovered, it was thought the universe could be explained in terms of a macroscopic projection of this. Some microscopic and macroscopic constants were found. The laws of physics are not so simple. The danger is that scientists might build upon a wrong assumption. What could happen if someone found an exception to Einstein's formula $E=MC^2$, for example, more energy was being produced than could be accounted for by total mass. Particle physics are now better understood thanks to particle acceleration experiments in places like the CERN tunnel. This has led to modern inventions. Scientists always want to explain away their findings. Sometimes the findings are so bizarre it opens up a new trend of thought. Progress is usually made step by step, building up information, opening up new avenues of approach. Clever is the man who can jump three steps ahead of his time as Albert Einstein. Some of the forces of nature are too great to reproduce on earth, so we can only observe and hope our laws of physics will explain.

Perhaps things are not totally explainable in scientific terms in order to reproduce the same experiment and get the same answer each time. We need new methods of analysis, big and better computers to analyse the data, and sometimes a different approach to a problem – called lateral thinking.

Spectroscopy is an important astronomer's tool for analysing the composition of matter and distance measuring. A heated substance will emit light, for example, light from a star. Light is made up of many colours of the rainbow. Pass light through a prism and a spectrum is produced.

Spectroscopy is the analysis of the wavelength dependent properties of light. Different elements produce different emission spectra. If light passes

4

LIGHT

$\longrightarrow$

Red
Yellow
Green
Blue
Violet

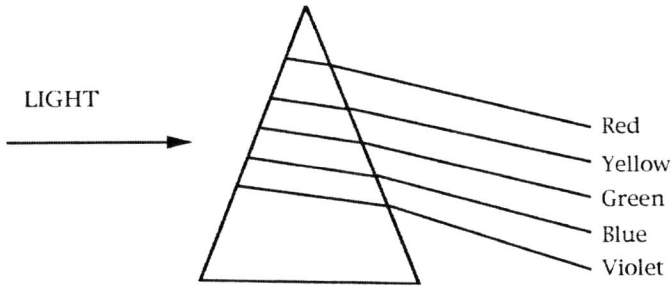

through a gas cloud, an absorption spectrum of dark lines shows which elements are present. Visible light is only a narrow band of radiation in the electromagnetic spectrum, which ranges from gamma rays to radio waves.

Gamma rays. X-rays. Ultraviolet. Visible light. Infrared. Radio Waves
TV   AM
Maritime
FM   Radio
Navigation

Spectroscopic analysis of emission and absorption lines is a very important part of astronomy and shows the abundance of elements in stars like our sun and universe. Everything that is made is made inside a star, all the elements that make us. The properties of light and radiation in the electromagnetic spectrum is a major study of the greatest importance. When nuclear explosions are detonated gamma rays are produced and can be detected. In 1967 the US launched a spy satellite to detect gamma rays in case the Soviet Union or another country exploded a nuclear device. What they actually detected was something quite extraordinary. A massive recording of gamma rays indicated something very strange was going on in the universe. In 1991 NASA launched a further satellite to study this effect. Bursts of unbelievable energy came from all over the sky, not from our own galaxy. On 22 February 2001 a very powerful gamma ray burst was detected and identified as a narrow beam of intense gamma rays from a black hole at very great distances, perhaps at the edge of the universe.

Lets talk about black holes. They exist; they have immense gravitational attraction and suck everything in, including light. What do we know about them? Not much; but we can still talk about them. Will we all become swallowed up by black holes which wander out from the centre of our galaxies? Do galaxies disappear into black holes forever, never to re-appear? What happens if several wandering black holes meet each other? Does one say to the other 'I am much bigger than you, so I am going to swallow you?' If they are identical in size and everything else (whatever that means), do they stay alongside like a loving couple? If they do try to join we may have a problem. If you sit on a stool that rotates and spin yourself round with your arms outstretched, and now bring your arms in, you rotate much faster. Black holes are spherical. If not the rotational speeds will be different at different places. Our earth has an equatorial bulge so the North to South Pole distance is less than equatorial distance by about 24 miles. Rotational forces are different at different places. If one part of a black hole is rotating close to the speed of light, another part may be rotating at more than the speed of light, which is not possible. So the black hole blows off energy in the form of gamma rays, through the centre of the black hole. The known laws of physics may change in black holes. Unfortunately if you venture into one you cannot send your findings out. Black holes can spin clockwise or anticlockwise. Some black holes have large electrical charges, some do not. Some are larger than others. Some emit radiation in gamma rays. They will vary in their rotational forces. The biggest black holes may be near the centre of our galaxy where the biggest stars implode. The overall amount of matter in the universe appears to be fairly constant, so if black holes are swallowing matter up, then matter must be created. I believe the energy source from them leads to the birth of a new star. Their presence will be spotted by the gravitational attraction of bodies close by, alteration of orbits, and eventually matter is seen to be sucked into the black hole. If you could travel near one and orbit it at great speed so you were not 'sucked in', time dilation would occur, one year in a spacecraft could represent 100,000 years on earth. Time is relative to the environment you live in.

Time on earth is so short. Perhaps that is why everything happens so quickly. The butterfly works so hard to reach maturity and fly about, saying 'look at me', I am so beautiful only to last but a short time. This enormous urgency of the life force is explained away as survival and reproduction. Is this the complete answer? Is it not related to the environment on earth with small gravitational forces? To overcome the earth's gravitational force in a rocket, we only have to leave the earth at 18,000 miles per hour to escape. If we go to the moon where gravitational forces are much less, we can jump much higher. The escape velocity to leave the moon's surface is much less. There is a balance between centrifugal force produced by rotation and gravitational force. So where does time come into the equation?

In the universe everything that is born, dies and is reborn. Creation and destruction go hand in hand. Mass and energy are interchangeable. Out of the fireball in the sky, after millions of years of cooling, come the beauty and diversity of creation. Quite extraordinary, that in the depths of the ocean some fishes have search lights, throw light bombs out behind them to avoid capture, make themselves nearly invisible, a star wars in the depths of the ocean. Is the earth an experimental greenhouse for growing things? What do we do if we have a greenhouse? We inspect it from time to time to see how things are growing, and maybe eat the fruits, or pick the flowers.

Space travel. Do spacemen come and visit our greenhouse? If life is in abundance throughout the universe it is reasonable to have had some visits. For us to visit them is beset with problems. If the nearest inhabited planet is in Alpha Centauri, then the distance is two billion, one hundred and eight thousand, seven hundred and thirty nine million, eight hundred and eighty thousand miles away. We are not going to do it in a rubber duck. If space is curved around objects the shortest distance across a curved surface is part of a curve. Recent research suggests that it is possible to bend light in a refracting medium under extreme conditions of electromagnetic fields by an enormous amount. Everyone has seen this bending effect when you dip a pencil into water. Does this reflective index apply to space?

Magnetism bends light.

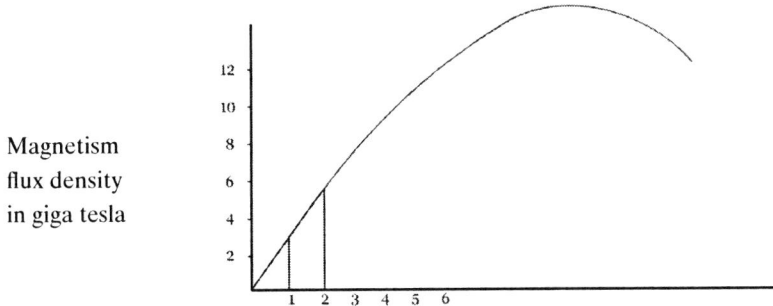

Magnetism
flux density
in giga tesla

A new science is developing called crossing superspace by 'Wormholes' which is very exciting (more about this in Chapter 2).

Travelling away from earth in a spacecraft at high speeds presents a time problem in returning home. Due to the difference in time between the people in a spacecraft and earth, two years spent in a spacecraft at high

7

speeds could see the space travellers coming home to an earth that has aged thousands of years. We must cross space in days, not years. Time dilation must occur to space travellers coming to visit us. Can we piece together a story of how we came to be?

Summary

Sixteen billion years ago, a big bang occurred. There was no time before this. In the beginning there was energy. What was it? Gamma rays? The building blocks of life were made from neutrons and protons. Photons of light were made. Slowly atoms were made, hydrogen in large amounts and carbon, nitrogen, and oxygen atoms. The build up of matter and big gravitational forces condensed the nuclei of the atoms and compression produces large heat, setting off a nuclear reaction, when hydrogen atoms were converted to helium with release of energy, a star was borne. There is enough hydrogen in the sun to keep the proton-proton chain going for millions of years. Everything we are made of has come from a star. DNA has made us, evolution has evolved us, earth's catastrophes have changed us, and eventually the sun will expand into a red giant and absorb us. Nothing is for ever. Change is normal. Black holes mop up the debris and when their rotational forces reach the speed of light they convert mass into energy to build new stars.

There are too Many Unanswered Questions

There are laws governing the universe. Who made the laws, the mathematics, the physics, the chemistry and everything else we have not yet discovered? The answer is a superior being to us. I find in nature and in the universe everything tends to return to a stable form, 'the neutral form'. The common denominator is the beauty and diversity of creation. Add to this 'how things are made from DNA', a planet with water, and life is in abundance in the universe. Cloud gases contain the chemicals that make DNA. These have been present billions of years before planet earth was made. What other forces exist in the universe we do not know about? There is not enough mass in the universe for gravitational forces alone to hold our galaxies together. Where is this dark matter?

I will try to answer some of these questions.

The Hubble Constant

Distant galaxies are receding from us at an ever-increasing rate. The further the galaxy, the greater the velocity of recession. This is measured by red

8

shifts, the more distant the galaxy the greater is its red shift, that is the shift of visible spectrum lines to longer wave lengths. The Hubble constant is continuously being altered. The present findings indicate a value of 73 kilometres per second per mega parsec (+ or − 7%). A mega parsec is equal to 3.1 x 10^{24} cm. The light from a distant galaxy takes longer to reach us so we are looking backward in time.

Another great discovery was made in 1965. Radio emission from the sky reaches us in equal amounts from all directions. Planck's law tells us the intensity of emission as a function of wavelength is related to temperature. Over the wavelength range of 0.2 to 21 centimetres the radiation approximates thermal emission at about 3°K. This gives us valuable information to feed into a cosmological model to try to predict the universe and its size. Can we feed in the right information when on earth we cannot simulate massive gravity, temperature, pressure, speed and other forms of energy? Scientific discovery is usually a slow process, step by step, approaching the problem from different angles. To further a new discovery we may have to build better equipment to obtain better results.

Measurements indicate the universe is expanding and the rate of acceleration is increasing in certain very distant parts. This may not be universal. We don't know the overall shape of the universe. Once the universe finishes expanding, it may remain in a steady state or contract. The total amount of matter in the universe shows no sign of change. It would be interesting to understand the density and weight distribution of matter in our own galaxy. Knowing the angular momentum (see Chapter 2, page 30) the various gravitation fields could be analysed. If our galaxy's rotation were to slow down and the angular momentum was reduced, it could be assumed matter would fall toward the centre, where the concentration of stars and black holes are greatest. A higher density galaxy will undergo more mergers. In lightweight galaxies a supernova explosion could blow away gas clouds reducing star formation. When galaxies collide or come very close to each other what happens to gravitational predictions, or are they thrown into a state of random chaos? Does the universe have a galactic centre about which it rotates? Do the most distant galaxies with a high velocity of recession act irrationally outside the known laws of physics? Do we know all the laws of physics? Are the known laws of gravity warped by space? We are constantly asking questions. Toward the edge of the universe, where matter is rarefied and gravitational forces are weak, or non-existent, the rate of expansion is increasing. There is 'no holding back'. Is this matter different to matter we know? The energy is different, not the matter. According to Einstein this is the dark energy making up the cosmological constant. This dark energy opposes the force of gravity. As the whole universe expands everything is moving away leaving less density. There is no evidence the expansion follows any symmetry. The rate of expansion has varied since the beginning. In the principles of general relativity only curvature and deceleration are connected. The speed of light has decreased since the big bang. The difference between theoretical physics and practical observation presents the biggest problem to understand. If the universe continues to expand at an increasing rate, matter will become rarefied, and temperatures will fall to a cold uninhabitable universe.

The biggest problem to solve is the total content of the universe and how it is made up.

Chemical elements made from stars and dark matter including radiation only account for 30% of the content of the universe. Where is the other 70%? Recent researches from many different viewpoints indicate the universe is full of energy of unknown origin, accounting for 70%. This is dark energy which does not absorb or emit light energy. It has negative pressure, and keeps everything apart opposing the forces of gravity. It in now known as a quantum field and given the name Quintessence. Is this an energy force derived from massive frictional forces opposing rotation?

In my introduction 'questions and answers', I quote 'can the universe totally implode'? The answer is no. I explained a new word 'spacuum', a space not measured by time and distance for the universe to expand into. Spacuum contains dark energy which repels gravity. It may be accelerating the expansion of the universe where matter is present at the periphery or retarding the expansion of the universe where matter is not present at the periphery. If the universal amount of matter present is constant, the universe may neither be expanding nor contracting. Perhaps the spacuum is receiving ejection of redundant matter and energy from a system that has now stabilised (see chapter 2 on spacuum).

References for further reading
1 Scientific American Dept Time BV. 415 Maddison Avenue, New York, NY 10017-1111; USA
2 Lawrence M Krauss, Ambrose Swasey Professor of Physics and Professor of Astronomy, Chairman, Department of Physics.
http://wwww.phys.cwru.edu/facultv/?krauss

Chapter 2

The formation of the earth, early plant and animal life, the chemistry and physics of life

Five billion years ago our solar system started to form from a cloud of gas in space. Rotational forces and gravity produced enormous forces and condensed the cloud gas until nuclear reactions started, and our sun was formed. Further debris consolidated to form planets at various distances from the sun. Heavier elements, iron and nickel fell to the centre of the earth and the fireball took a long time to cool. In another 1,000 million years time the earth's crust started to form floating on the molten rock layer below the surface. It is still cooling today. There are 1000 volcanoes which are active at different times, pouring out lava from inside the earth. Some of the lava comes from the deepest part of the earth, the origins of the earth 4,500 million years ago. 3,000 million years ago sufficient solid crust formed to create land, but even today the earth's crust is only four miles thick in places like the pacific ocean. In other areas it is 40 miles thick. The surface of the earth is cooling. The interior is being heated by radioactive decay. We are fortunate that our earth has volcanoes to vent the heat locally, like safety valves. The whole surface of some planets is boiling.

As the planets formed some are much bigger than others. They all orbit the sun from different distances. The earth has a diameter of about 8,000 miles, Jupiter 88,000 miles, Saturn 75,000 miles and the sun 865,000 miles.

The volcanoes erupting produced oxygen and hydrogen to make water. The temperature became cool enough for the ocean not to boil away. We are at a critical distance from our sun at a time of earth's cooling, so life can form. In another 1000 million years time the sun will expand into a red giant, the oceans will boil away, life on earth will cease, time to leave for another planet.

Orbiting Times for Planets to Orbit the Sun

Mercury	Venus	Earth	Mars	Jupiter	Saturn	Uranus	Neptune	Pluto
88 days	225 days	365 days	687 days	11.86 years	29.46 years	84 years	165 years	248 years

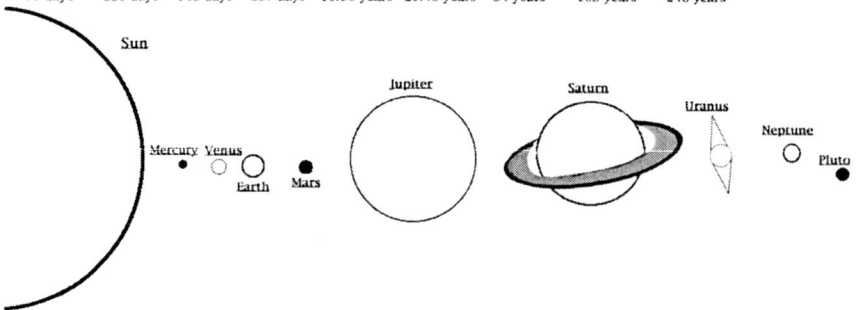

Mapping the Earth Today

The earth is not a perfect sphere. When celestial bodies form, heavy elements sink to the centre like nickel and iron, and this combined with a centrifugal force, causes an equatorial bulge to develop. The equatorial bulge of the earth is about 24 miles greater than the North/South polar distance. The earth is tilted on its axis and it orbits the sun in 365 days. Different planets take different times to orbit the sun and at different speeds. The earth is travelling at 18.2 miles per second or 66,500 miles per hour.

The geographical axis of the earth is a line through the North and South Poles about which the earth rotates. The earth rotates from West to East completing a 360 degree turn in 24 hours.

The equator divides the earth into two halves north and south. Latitude is a measurement of how far north or south you are from the equator. It is measured in degrees from 0 at the equator to 90° at the poles. Latitude is a measure of distance. One degree is divided into 60 parts, each part is called a minute.

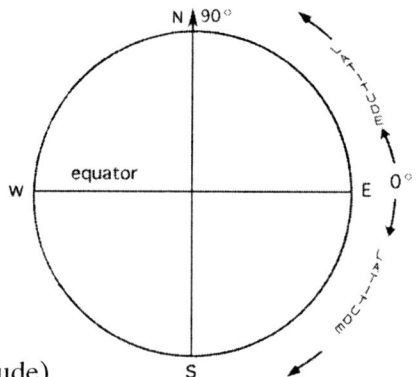

1 minute = 1 mile (adjacent to the latitude)

12

Longitude is an angular measurement from zero longitude, which is called the Greenwich Meridian, passing through London. It is measured 0–180° to the west from the Greenwich Meridian, and 0–180° to the east, meeting in the Pacific in the International Dateline. Geographically lines of longitude curve and meet at the poles.

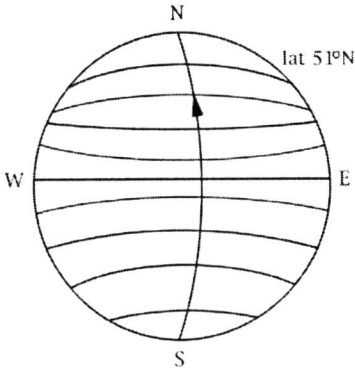

England is at latitude 51° north, longitude 0° west, or 51° x 60 minutes = 3060 miles north of the equator. The shortest distance across a curved surface is part of a curve. It is not practical for a sailor to sail a curve, so mariners use Mercator charts invented in 1600.

The Mercator Chart
We cannot draw a curved surface on a flat piece of paper. We can take a small part of the earth's surface and draw it without distortion.

Constructing the Mercator Chart
1 Lines of latitude and longitude are drawn to cross at a right angle.
2 We have to expand the latitude scale to compensate for the longitude distortion.
3 As a result of this, land masses appear larger at higher latitudes, but angles and distances are correctly represented.
4 We can draw a straight line called a rhumb line between any two positions and measure the distance against the latitude scale. Mariners can measure angles and can automatically get nautical miles, but they cannot measure distance accurately the way a land surveyor does. An average sea mile is 6,080 feet or 1.87 km, a land mile 5,280 feet or 1.61 km.

There are other projections which map out the earth in a grid form so any position on earth can be identified. There are satellites orbiting the earth which can locate any position on earth to within 30 feet. Navigators at sea have a receiver called GPS = Global Positioning System, which reads out their position in latitude and longitude.

13

Cherbourg } latitude 49° 41·00N
longitude 01° 35·70W

The Distance to go from the Isle of Wight to Cherbourg is 60 miles. The true course can be found from the compass rose.

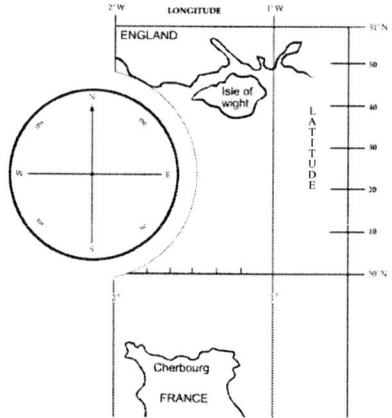

The Magnetic Effect of the Earth
The geographical axis of the earth is true north and south, and is not the same as magnetic north and south. The earth acts like a bar magnet. The iron core is too hot to be magnetic but the rotation of the core, slightly slower than the surface causes electro dynamic forces which create a magnetic field. This attracts a freely suspended magnetic needle causing the north-seeking pole to be in Hudson Bay and the south-seeking pole in Australia. The direction the magnetic needle points is called Magnetic North. It varies with your position on the earth's surface, and from year to year, as the earth's magnetism changes. You can buy a yearly Isogonic chart which shows lines of equal variation, and this enables the navigator to obtain the difference between true and magnetic north called Variation.

If from where you are, the magnetic north points to the left of true north, variation is called West. If from where you are the magnetic north points to the right of true north, variation is called East.

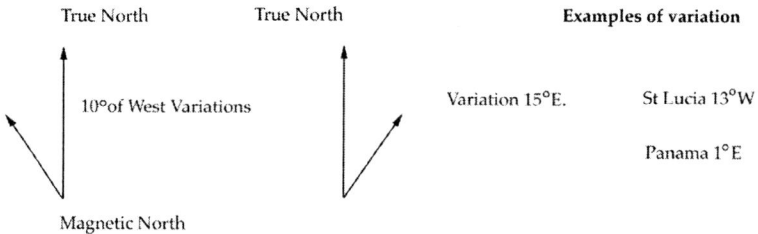

True North   True North   Examples of variation

10°of West Variations   Variation 15°E.   St Lucia 13°W

Panama 1°E

Magnetic North

A sailor uses a compass to navigate his ship. Birds, homing pigeons, turtles and whales have built in tiny magnets in their brain cells, allowing them to navigate by magnetic lines of force.

The earth's magnetic field changes, magnetic north getting closer to true north in only a matter of 100 years, and then moving away again. This is due to the earth's orbital wobble on its axis, and in the course of many thousand of years total polarity change can occur, when the north seeking compass needle points south to Australia. There is evidence in rock strata in Australia this could have happened 25,000 years ago, but scientists disagree over the date, the more accepted date being 200,000 years ago. It has been forecast to happen again within the next 200 years. This will cause worldwide problems to man, animals, birds and sea life; also to the Van Allen Belts (described below).

THE MAGNETOSPHERE (OR MAGNETO SHEATH)

The effect of the earth's magnetic field extends about 40,000 miles into space and is called the magnetosphere. It protects the earth from damaging radiation particles from the sun. Every so often the sun emits extra energy as solar flares. This compresses the magnetosphere and the solar wind is defected and trapped as atomic particles in layers described as Van Allen Belts.

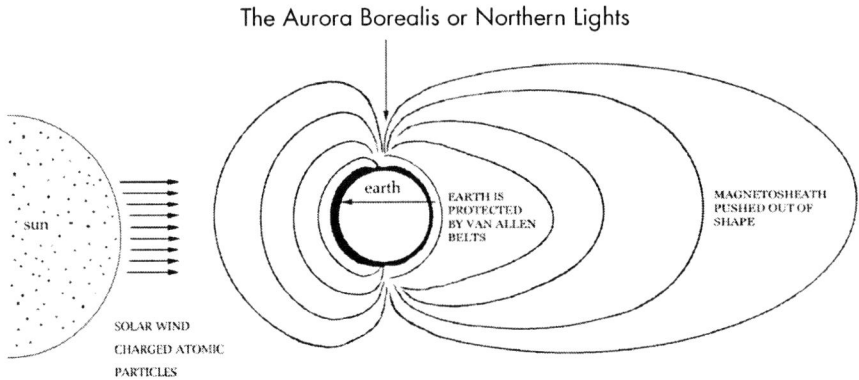

The Aurora Borealis or Northern Lights

sun

SOLAR WIND
CHARGED ATOMIC
PARTICLES

earth

EARTH IS
PROTECTED
BY VAN ALLEN
BELTS

MAGNETOSHEATH
PUSHED OUT OF
SHAPE

The Moon

Debris trapped by the earth's gravity caused a planet to form orbiting earth. other planets have moons. Jupiter has 16 moons. The elliptical orbit of the moon is so predictable, the gravitational pull on the ocean causing tides can be predicted years ahead. Over many millions of years the moon's orbit will change and eventually the moon will come closer to earth and then fly off into space. The moon's orbit is elliptical and its distance from the earth varies throughout the month. When it is nearest to the earth the moon is said to be in PERIGEE, and furthest away in APOGEE. One complete orbit takes 27½ days. Higher tides called Spring tides occur every two weeks when the moon is closest to the earth and smaller tides called Neap tides occur when the moon is furthest away. The biggest tides occur twice a year when the sun and moon are in line exerting greatest force. This occurs at the equinoxes, September 21st and March 21st.

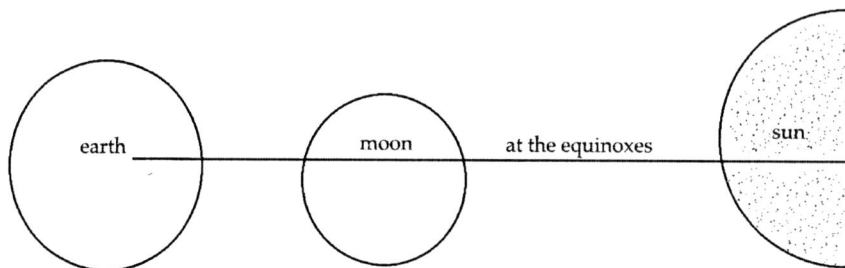

The tide is the vertical rise of water. The tidal stream is the horizontal motion produced by the rise and fall. The rising tide is called the Flood tide, the falling tide the Ebb tide. Ocean currents occur due to permanent wind direction, trade winds, which cause friction at the surface. The moon exerts the same pull on different parts of the earth's ocean, but it cannot raise the water where it is very deep. The tide is a shallow water effect, where estuaries are shelving. An example of a large rise in tide is the Bristol Channel in England. On spring tides the rise is 40 feet, which means that twice a day (the flood tide is 6 hours, the ebb tide 6 hours) a vessel on a mooring which dries out at low water spring tides will rise 40 feet at the top of the tide. In offshore areas, eg East Coast of America, West Coast of Portugal and Spain, Spitsbergen latitude 80°N, the spring rise is about 12 feet (see page 21 for feet/metres conversion).

The Earth's Crust
As the earth cooled so a crust of rocks formed a large continent. The volcanoes formed our atmosphere and also produced water vapour to make the oceans. 70% of the earth is covered by water. In years of big volcanic activity fine dust is deposited in the atmosphere, for example, Krakatoa erupted in 1873 and in 535AD. This blots out the sun and lowers earth temperature by as much as 1½°; Ice builds upon ice which thickens the ice

over the polar cap. Cold water from the melting ice causes warmer currents, eg the Gulf Stream in the Atlantic Ocean, to drift south causing 'ice ages'. In 15,000 BC ice reached the South of France

Volcanic activity pours lava from molten chambers deep inside the earth, not only on land but also under the sea. The earth's crust is only 4 miles thick in places under the Pacific Ocean, up to 50 miles thick in other places, like the Himalayan mountains. There is a lot of volcanic activity under the Pacific Ocean. If the water temperature rises by only 2 degrees, the air above will be heated, and this hotter water/air will drift across the Pacific to the west coast of America and into the Caribbean and Atlantic ocean. The hotter air combining with cold polar air will produce more violent storms. If a basalt lava flow 25 miles long came up under the Pacific Ocean, the rise in temperature could melt the icecaps with a rise in sea level of 400 feet. There is concern at present about the 'Greenhouse' effect, that man is polluting the atmosphere causing the earth to heat up. The history of the earth is going to show periods of global warming and global cooling.

Volcanoes produce new islands and new land, erosion wears away the land. Rocks forming the crust are moving in plates. Some plates are undercutting other plates, some building mountain ranges like the Himalayas, and associated with this movement at the plate boundary are earthquakes and volcanoes. There are maps in textbooks showing the distribution of the earth's 15 large plates, where some plates are moving together and some plates are moving apart. San Francisco lies on a plate boundary and people live in fear of a major earthquake. The continental plates cause land masses to drift apart. 2000 million years ago one large continent started to break up. The Atlantic Ocean has developed from the drifting apart of the continents of Africa and America.

THERE ARE BOOKS DEVOTED TO PLATE TECTONICS

Volcanoes

There are 1,000 active volcanoes. Many volcanoes and earthquakes occur near plate boundaries, where the crust movement allows magna to rise up from deep inside the earth. Many are under the sea (heating the oceans). When a volcano formed a new island off Iceland in 1966, vegetation started to grow in 20 years, illustrating the fertile nature of volcanic soil, especially soaked in the right medium of ocean nutrients and a good rainfall. When a volcano erupted on Lanzarote nearly 300 years ago, it sucked up salt seawater, making it impossible for vegetation to grow due to poor rainfall

There are books devoted to plate TECTONICS

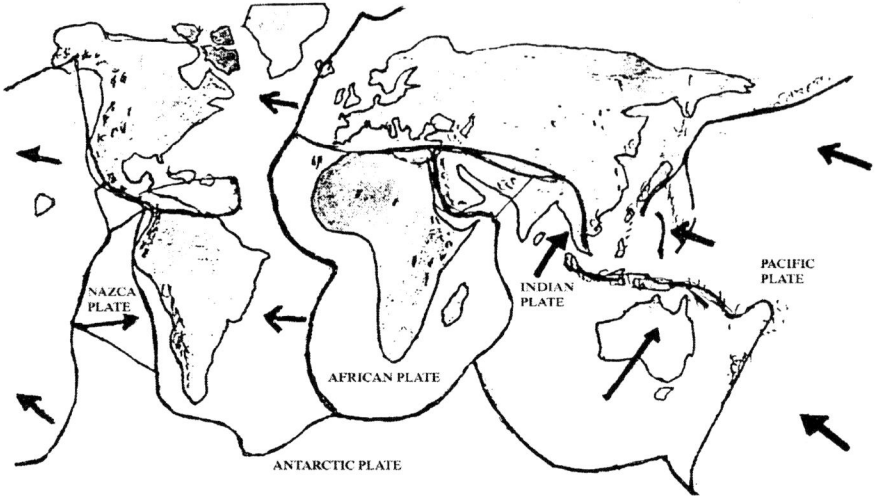

in Lanzarote to wash out the salts. Water is an essential requirement for living things.

Pyroclastic Flow

A pyroclastic flow may follow an eruption. This is a cloud of very hot gas, ash, dust and lava, which is so thick that it flows like water down hill, across towns and countryside at incredible speeds (300 mph) and very high temperatures (700°C). An example was the pyroclastic flow from Mount Pelee in the Caribbean Island of Martinique in 1902 which killed 39,000 people in three minutes.

Famous Volcanoes

In recent times most people were killed in the following eruptions: Tambura 1815, 90,000 killed; Mount Pelee, 1902, 39,000 killed; Krakatoa, 1883, 36,000; Nevado del Ruiz 1985, 23,000; Santorini 1470 BC, Vesuvius AD 79 and 1944, Taupo 186 AD, Krakatoa 535AD 1680, 1883, Towada (Japan) 915, Oraefajukuu (Iceland),136, Mount Pinatuba 1380AD, Mount Etna (Italy) 1669, 1947 and 1991, 1680, Hekla (Iceland), 1693 and 1970, Stromboli (Italy), 1768 and 1989, Laki Fissure 1783, Tambora (Indonesia), Mount St Helens (USA), 1800, 1887, 1980, 1989, Kilauea (Hawaii) 1823 and 199, El Pelée Martinique) 1902, Santa Maria (Guatemala) 1902, Mount Katmai (USA) 1912, Bezyhianny 1956, Helgafell, Monserrat (Caribbean) 1997.

Volcanoes Under the Sea

These volcanoes have four effects (1) building new islands, (2) increasing the water temperature, Pacific and mid Atlantic ridge. (3) Tsunami (seismic waves), (4) depositing rich minerals – (volcanic soil which is good for growing crops).

Earthquakes

Seismologists study earthquakes. They are measured on the Richter scale. A severe earthquake in 1906 destroyed San Francisco USA. In 1994 Los Angeles USA an earthquake caused devastation. 80% of earthquakes occur in the Pacific and North East Coast of Asia. Earthquakes can cause landslides, mud flows and tsunamis. The Richter scale measures the amount of energy released, on a scale from 0–8 and over 8 is serious destruction. More people die from earthquakes than from volcanoes. In the last 100 years more than one million people died from earthquakes. A seismometer records the shakes during an earthquake, with a pen recording the movement of a rotating drum.

Erosion

Two thirds of the earth's surface is covered by water. The wind and rain act on the rocks producing erosion and soil. This is called weathering. In wet climates rain attacks the minerals in the rock; chemical weathering. In dry climates, deserts, sand blown by the wind attacks the rocks; physical weathering. Mountain rivers wear away softer rocks, leaving harder rocks behind. The burning of fossil fuels can produce sulphur dioxide, a gas which in contact with water makes acid rain. This can damage buildings, trees, pond and lake life. A soil profile is a slice of soil down to the bedrock. The soil from deserts contains hardly any organic material.

Examples of Adaptation to the Environment

Lizards have had 300 million years to adapt. The Draco Dragon lives entirely in the trees. With their four feet they can grip the bark and climb to the top of a tree. When danger presents itself, they can hold their loose skin open and make a wing to glide from tree to tree. They signal to each other by displaying coloured stretches of skin which looks like semaphore. They practice the art of hibernating in the trees. The dragon lizard can adapt to the floor of the forest after a forest fire. They take advantage of the lack of shelter for surviving insects which makes for easy pickings. They have a problem to hide from birds. To ward off an attack they arch their skin over

Fossils are studied by Palaeontologists. They are the remains of plants and animals that lived millions of years ago, and have become preserved in rock. Approximate times life appeared on earth

Million years ago	Description	Illustration of each development era
2000 million	Single cell organisms in the sea	
1000 million	Plankton forming in the sea	
700 million	Corals and jellyfish	
600 million	Crustations in the sea	
500 million	Fish appeared in the sea	
400 million	Land plants	
380 million	Insects	
340 million	Frog, amphibian, crocodiles	
300 million	Reptiles, turtles, lizards, snakes	
250 million	Dinosaurs	
225 million	Mammals, mice, small shrewlike animals (protothercan)	
150 million	Birds	
140 million	Flowering plantation	
350 million	First apes	
2 million	First human fossils found in Africa	

Time periods in biology
TRIASSIC   248–213 million years ago
JURASSIC   213–144 million years ago
CRETACEOUS   144–65 million years ago
65 million years ago Dinosaurs became extinct.

their heads and put on an aggressive appearance to challenge their opponents.

Lizards also live in the Australian scorching interior. They absorb moisture through their feet, and there is enough desert dew to do this. They

20

eat ants. Their grotesque appearance gives them the name Thorny devils. Another lizard lives in caves to escape the sun. They are called Rock dragons. Although small in size, they can outrun any other lizard. Raising their front legs they run on their back legs which operate alternately producing a bicycle like effect. They can reach speeds of 20 mph and outrun the very large, powerful, and aggressive lizard, the Monitor.

When Krakatoa erupted in August 1883, seismic waves were generated called tsunamis, some 100 feet high. They wiped out civilisation and nature in low lying neighbouring areas, and the final devastation was caused by the waters retreating from the flooded areas. Some Java rhinos escaped on higher ground. How did nature repair the damage? Ujung Kulon is a national park about 25 miles from Krakatoa, where man has not returned to live for over 100 years. David Attenborough took a boat up the Geegenta River to see how well nature has regenerated the environment. In his words 'it is like going back to prehistoric times'. The success story is due to the climate, 100 inches of rain for 6 months followed by six months of sunshine. Adaptation to this environment is centred around escaping predators, finding food and finding mates. All vegetation was buried under volcanic dust and ash which litters the mud banks of the river. Crocodiles are numerous. The mudskipper does not know whether he wants to live on land or in the water, and he skips across the water. A fish swims upside down holding a leaf above himself to hide from birds. This adaptation has never been seen before in nature. Another insect can walk on the water. The surface tension of its feet on the water does not break the surface tension of the water. All this diversity of creation, and much more, has occurred in only 100 years!

Feet/metres Conversion

Feet		Metres	Feet		Metres	Feet		Metres	Feet		Metres
3·28	1	0·30	45·93	14	4·27	88·58	27	8·23	131·23	40	12·19
6·56	2	0·61	49·21	15	4·57	91·86	28	8·53	134·51	41	12·50
9·84	3	0·91	52·49	16	4·88	95·14	29	8·84	137·80	42	12·80
13·12	4	1·22	55·77	17	5·18	98·43	30	9·14	141·08	43	13·11
16·40	5	1·52	59·06	18	5·49	101·71	31	9·45	144·36	44	13·41
19·69	6	1·83	62·34	19	5·79	104·99	32	9·75	147·64	45	13·72
22·97	7	2·13	65·62	20	6·10	108·27	33	10·06	150·92	46	14·02
26·25	8	2·44	68·90	21	6·40	111·55	34	10·36	154·20	47	14·33
29·53	9	2·74	72·18	22	6·71	114·83	35	10·67	157·48	48	14·63
32·81	10	3·05	75·46	23	7·01	118·11	36	10·97	160·76	49	14·94
36·09	11	3·35	78·74	24	7·32	121·39	37	11·28	164·04	50	15·24
39·37	12	3·66	82·02	25	7·62	124·67	38	11·58			
42·65	13	3·96	85·30	26	7·92	127·95	39	11·89			

COMPOSITION OF THE EARTH

Everything is made inside a star, in our case the sun. 97% of the mass of the sun is made from hydrogen and helium. All the elements come from the sun. The majority of these elements are metals. 112 elements have been identified.

Examples are: Silver, aluminium, gold, carbon, calcium, chromium, copper, iron, hydrogen, helium, mercury, iodine, potassium, magnesium, nitrogen, sodium, nickel, oxygen, phosphorus, lead, platinum, sulphur, silicon, tin, titanium, uranium, tungsten, zinc.

The earth is made up of:

1	OXYGEN	46%
2	SILICON	28%
3	ALUMINIUM	8%
4	IRON	5%
5	CALCIUM	3.6%
6	SODIUM, POTASSIUM, MAGNESIUM	7.5%
7	OTHER ELEMENTS	2.0%

75% of the earth is made from oxygen and silicon. The crust of the earth is made of light elements chiefly oxygen, silicon and aluminium. The continental rock is chiefly granite. Ocean rock is mostly basalt, eg The Giants Causeway Northern Ireland. Granite and basalt originate from molten magna. Granite is a mixture of Feldspar, Quartz and Mica.

Composition of the Atmosphere

The Troposphere is the first layer up to 12 miles high that contains 80% of all the gases and it is where our weather is formed. The gases are nitrogen 78%, oxygen 21%, other gases 1%. We need oxygen. Nitrogen is just breathed in and out. As we move upwards temperature starts to fall with height. The weight of all the atmosphere pushes downwards and is called atmospheric pressure. Other gases we breathe in, in the one percent are;

1	Argon	5	Methane		
2	Carbon Dioxide	6	Krypton		
3	Neon	7	Hydrogen	9	Zenon
4	Helium	8	Ozone	10	Water vapour, dust, industrial pollutants

The STRATOSPHERE is the next layer, 12–30 miles high. It contains the ozone layer, a form of oxygen that protects us from the sun's harmful rays. Temperatures are very cold – Aeroplanes fly in the calm area above the weather, eg Concord. There is a hole in the ozone layer the size of North America, over Antarctica. This is letting in dangerous radiation, ultraviolet rays, from the sun. Pollutant gases could be causing this effect and causing global warming. Carbon dioxide is a 'greenhouse' gas that may be preventing heat escaping from the surface of the earth so the temperature is rising globally.

The MESOSPHERE is the next layer, 30–50miles high. The atmosphere is very cold and rarefied. Meteors burn up in this area producing 'Shooting Stars'. The highest weather balloons reach the mesosphere.

The THERMOSPHERE is the next layer 50–280 miles high. It has a very rarefied atmosphere and contains electrically charged air that reflects radio waves in the 'Ionosphere'. The Aurora borealis or 'Northern lights' appear in this layer.

The EXOSPHERE is very rarefied and spacecraft leave the earth's atmosphere at this level. So most atmosphere is in the first 200 miles

MATTER

Matter exists in the form of solids, liquids and gases. Matter is measured in terms of its mass, density and volume. Mass is how much matter exists in an object. Volume is how much space it occupies.

DENSITY is $\frac{mass}{volume}$ so density compares the heaviness of different materials.

Gold is very dense. Steel is less that half as dense as gold. Concrete is only a third as dense as steel.

ATOMS

These very tiny particles make the building blocks of the universe, Atomic structure – the nucleus contains protons and neutrons. Electrons orbit the nucleus in layers. Protons have a positive charge. Neutrons have no charge. Electrons are negatively charged particles. The total number of protons and neutrons in the nucleus is the atom nucleon number.

Hydrogen atom

One electron
One proton

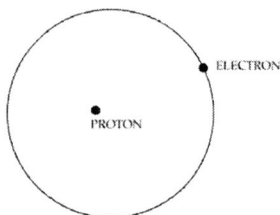

ELECTRON

PROTON

Different elements have
different numbers of protons
neutrons and electrons.
Atoms contain smaller units
called subatomic particles.

Carbon 12 atom

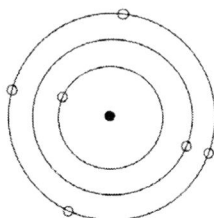

Nucleus contains:

6 Protons

6 Neutrons

Electrons

eg. Carbon has 6 protons and 6 neutrons called Carbon 12. There are 100 types of atoms. An element is made of only one type of atom. Most elements are metals. Alloys are a mixture of different metal e.g. Brass is copper 70% – Zinc 30% – Bronze is copper 70% – Tin 30% – Pewter is tin 73% lead 27%. Stainless steel iron 70% – Chromium 20% – Nickel 9.5% – Carbon 0.5%.

Potassium, sodium, calcium and magnesium are abundant metals making vital roles in earth's chemistry. We know calcium is present in bones, teeth etc.

NON-METALS
These are vital to life on earth eg sulphur, hydrogen, oxygen, carbon, nitrogen, chlorine, bromine, fluorine and iodine. The smell of sulphur is in volcanoes and sulphur springs and it is abundant in the universe. On Venus there are sulphur jets thrusting pure sulphur 60 miles into the sky.

A COMPOUND. A MIXTURE
Most substances are mixtures. A compound is made of two or more elements chemically bonded to make a new substance.

A combination of elements may produce a different substance eg sodium + chlorine = Sodium Chloride (common salt)

Carbon compounds make up most living things and are present in many everyday compounds.

CHEMICAL REACTIONS

New substances are formed and the atoms re-arrange themselves to form new substances. No mass is lost. Elements have a chemical symbol so a chemical equation can be written.

Examples of chemical formula:

oxygen is made of 2 atoms of oxygen	$= O_2$
water is 2 hydrogen atoms + 1 oxygen atom	$= H_2O$
carbon dioxide is 1 carbon atom + 2 oxygen atoms	$= CO_2$
hydrogen sulphide is 2 atoms of hydrogen + 1 of sulphur	$= H_2S$
methane is 1 carbon atom + 4 hydrogen atoms	$= CH_4$

Chemical reaction

copper + sulphuric acid = copper sulphate + hydrogen

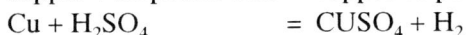

$Cu + H_2SO_4 \qquad = CUSO_4 + H_2$

Substances called Catalysts speed up the reaction but themselves remain unchanged by the chemical reaction.

Acids and Alkalis

Acids react with metals and can dissolve them

strong acids are;- Hydrochloric acid HCL

Nitric acid HNO_3

Sulphuric acid H_2SO_4

The strength of an acid or alkali is measured on a pH scale. 1 is highly acidic, 7 is neutral, 14 is highly alkaline. You can either use pH meters measuring the concentration of hydrogen ions, or pH paper or solution using a colour code.

BIOLOGY

The Carbon dioxide cycle

1 Carbon dioxide is present from animals and humans respiration and from volcanoes.

2 Plants absorb carbon dioxide.

3 Animals eat plants and dead animals are broken down by bacteria converting carbon back to carbon dioxide.

4 Fossil fuels, peat, can be used as fuel, which produces carbon dioxide.

The Oxygen cycle

1 Animals and humans breathe oxygen in the air.
2 Carbon dioxide is breathed out.
3 Carbon dioxide is absorbed by plants.
4 Plants release oxygen.

Protein and amino acids are the building blocks of life. The nitrogen and sulphur cycles help to make them.

The Sulphur Cycle

1 Plants absorb sulphates and make amino acids.
2 Plants are eaten by insects.
3 Decomposition produces hydrogen sulphide.
4 Bacteria use oxygen to make sulphates to replenish the soil.

The Nitrogen Cycle

1 Nitrogen is in abundance in the air and is used by plants to make protein.
2 Animals eat plant protein.
3 Body wastes are converted to ammonia and back to nitrates by bacteria and fungi.
4 Plants absorb nitrates. Fertiliser is artificial nitrates.

Photosynthesis

Plants made chlorophyll and can absorb carbon dioxide with water to make food and breathe out oxygen. Oxygen released by plant photosynthesis balances O_2 used by other species.

The Wonders of Nature and the Living World

There is such beauty in flowers and trees, in corals under the sea, in insects, in fishes and sea life, in birds in the sky, in animals on earth, that it never ceases to amaze me what I will read about and see next time I step out of doors. How is it all kept going? Sex is the answer. Flowers have male and female parts. Anthers are the male parts and they make tiny pollen grains. The ovary is the female part which makes eggs called ovules. Pollination occurs when pollen joins the ovules and the ovules make new seeds which are scattered in many ways.

Many plants are capable of self pollination ie they have both male and female organs. Cross pollination produces more genetic variations. Many flowering plants depend on animals for cross pollination. Bright colours and scents attract the animals and sticky pollen attaches itself to the insects and animals. An example of wind pollination is the Hazel Catkin. Flowers die and new ones replace them. Nature ensures continuous reproduction. I mention 5 ways in which seeds are scattered.

1 Wind – some seeds are very light and get blown in the wind eg, orchids.
2 Some seeds develop parachutes to catch the wind eg, thistles.
3 Some seeds are sticky and have hooks to catch on birds and animals.
4 Some fruits have seeds inside. Animals eat the fruits. The seeds are deposited in animal droppings.
5 Coconuts are seeds which float and are carried by the sea to grow into new palm trees.

The First Cells Were Formed in the Sea
When the 'cell to be' wrapped a membrane of cellulose around itself, this was the start of organisation nutrients coming in, waste products leaving. No nucleus or controlling centre was present. It was much later a nucleus was formed containing chromosomes. Protists cells developed a nucleus and flagella like threads to thrash the water to help absorb nutrients. These cells had a cell membrane', a nucleus (control centre) and a substance in the cell called cytoplasm in which early chemical reactions took place. Muhera is the name given to primitive one-celled organisms that do not have a nucleus. They take in nutrients through a cell membrane. Some make their own food by photosynthesis like plants using chlorophyll in chloroplasts. These are blue and green algae and early bacteria.
    Examples of protist are:-
        1 Euglena, 2 Amoeba, 3 Paramecium, 4 Diatoms

Euglena use sunlight to make food. Amoeba wrap themselves around food and absorb it. Fungi absorb food from their surroundings. At a later date these cells develop a stronger outer cell wall made of cellulose to give rigidity and protection to the cell. Animal cells developed specialised nerve cells, and creepers moving across the soil had a primitive nerve centre. When plants put down roots, these primitive nerve centres were absorbed The dead and living cells in the sea formed the plankton or soup on which corals, jelly fish and crustations developed.
    In 2002 a most remarkable series of television programmes were shown by David Attenborough about the deep ocean. Never before have cameras gone to such great depths in the ocean. The result is the most strange

looking creatures and a mini 'star wars' in these great depths. There is no light, only massive pressures. Some have developed very large eyes, transparent bodies, search lights and the 'art' of throwing light bombs out behind to avoid capture. We can see phosphorescence in the sea when tiny plankton emit light. Why do they do this? Some insects can do this. Fireflies are beautiful to watch and emit light to attract mates. Light energy is stored and slowly released by these organisms. Glow in the dark paints absorb light energy and emit it. There is no light in the deep ocean. How do these creatures store light if there is none? Do they make fluorescent chemicals in order to throw out light bombs? Does Darwin explain this? Darwin published his first book following a voyage of discovery on a sailing ship the Beagle to South America, 'A naturalists Voyage on the Beagle 1839'.His theory of evolution was written in two other books 'Origin of Species' 1859 and the 'Descent of Man' 1871. Darwin says, 'all things develop from very simple forms through natural selection among variations'. This favours individuals and species best suited to a given environment and survival of the fittest.

Back to my earlier question – was it chance that produced the first cell membrane, and cytoplasm, where primitive chemical reactions take place? Now I am convinced of a different explanation. A natural chemical attraction occurs first, the cell membrane came afterwards. Chemical substances can be placed in a list of those that react most, and those substances that react least. A catalyst can accelerate a reaction, but organic chemistry controls us. The natural tendency in nature which predominates is to return to the natural attraction between certain chemicals. This is its stable form.

NUCLEAR PHYSICS AND RADIOACTIVITY

Most atoms are stable. The nuclei of some atoms are radioactive and emit radiations called alpha, beta and gamma rays. These are dangerous radiations which can be measured with a GEIGER COUNTER. Most elements have radioactive forms called Radioisotopes. They are used in medicine for example, for measuring the uptake of radioactive iodine in the thyroid gland in the neck $C_{15} H_{11} O_4 NI_4$. Radiation is used for sterilizing equipment and in the treatment of cancer. Alpha rays are streams of positive charged particles made up of two neutrons and two protons. Alpha particles can be stopped by sheets of paper. Beta rays are electrons, more penetrating and can be stopped by a sheet of aluminium. Gamma rays are similar to X-rays. They are the most dangerous and penetrating type of radiation. They are electromagnetic waves which can be stopped by a lead screen.

Nuclear fission is the splitting of an atom's nucleus producing energy. The nucleus is made of protons and neutrons which are held together by the strong nuclear force. Nuclear fusion is the joining together of atoms to make larger atoms, eg when two isotopes of hydrogen, deuterium and tritium collide fusion takes place releasing a neutron. This larger atom is called Helium. These are the reactions which are happening in the sun. When the nucleus of an atom of uranium is struck by a neutron it causes fission and the result of splitting the atom produces energy as heat. The resulting atom is unstable and causes more uranium nuclei to split, resulting in a chain reaction. Uranium is a silver white metal, found in rocks, used for fuel in nuclear reactors and making atomic bombs. Plutonium is a radioactive metallic element mostly produced artificially. It is very explosive. Thorium is a radioactive metal which is plentiful in rocks.

Decay Series and Half Lives
Radioactive substances emitting radiation particles use up the mass of their nucleus. The rate is measured in terms of how long half of the original is used up or half of the atoms in a radioactive substance to decay. As the particles are emitted a different element will be formed until eventually a stable nucleus is achieved. For example, Uranium 238 has two stages of beta decay, five stages of alpha decay and others, eventually the element lead 206 is made. The half life of different substances varies enormously. The half life of those isotopes producing dangerous gamma rays are the shortest eg;- Cobalt 60, 5.3 years. Uranium 238 has a half life of 4500 million years, so there should be plenty of nuclear fuel left on earth for years to come. Uranium 235 is unstable above a critical size. If we take two blocks of sub-critical size and throw them together = detonate, we can produce a massive chain reaction, a nuclear bomb. If we can control the energy flow as in the early reactors using carbon rods, we can harness the energy in a nuclear power station. Everything is made in the sun. It is like going to the supermarket to do our shopping, except the sun is a bit too hot and things have not yet been put into little packages. Different planets at different distances from the sun are making a different soup.

Motion - Definitions
SPEED is the distance divided by the time taken.
ACCELERATION is increasing their speed.

VELOCITY is the rate at which an object changes its position. An object going round a corner at constant speed has a changing velocity.

FRICTION, like riding a bicycle causes your speed to fall due to mechanical parts rubbing together. Friction opposes motion and on earth stops things moving. In space there is no friction to motion, so objects can continue moving.

INERTIA, you need effort to change the state of motion, whether you are moving or at rest. MOMENTUM, is velocity multiplied by mass

CONSERVATION OF MOMENTUM, a snooker ball strikes a stationery ball. The momentum of the first ball moves the other ball. The total momentum of the balls is the same.

ANGULAR MOMENTUM, a gyroscope or a spinning top has angular momentum to resist the force of gravity trying to topple it. A gyrocompass is used for navigation and it is set in motion pointing towards north.

LINEAR motion is motion in a straight line

OSCILLATE, an object is defined as oscillating when it moves back and forwards about a fixed point eg an oscillating pendulum.

CENTRIFUGAL FORCE – circular motion – spin on a chair with your arms outstretched, now bring your arms in and you spin faster. This is the conservation of angular momentum. The magnitude of constant angular momentum is 1) body's mass, 2) speed, 3) radius of the orbit.

Isaac Newton (1643–1727) produced three laws of motion.

1. An object will remain at rest or move in a straight line until a force acts on it.
2. The rate of acceleration of a moving object is proportional to the force acting on it in the direction of the force.
3. When one object applies a force on another, the other object exerts an equal and opposite force on the first.

Gravity

Gravity is a force pulling one object toward another object. The mass of each object is involved. Allow a ball to drop to the ground. Because the earth mass is greater than the ball, so its gravitational pull is stronger. If we put the same mass on the moon, because the moon's gravity is only a sixth of the earth it weighs less. So the weight of the ball is mass x gravity.

Newton's law of gravitation

To find the force of gravity between two objects multiply their masses and divide by the square of the distance between them.

If the moon was two times its mass the force of gravity between the moon and the earth would be two times as great.

Diagram

Earth   Moon

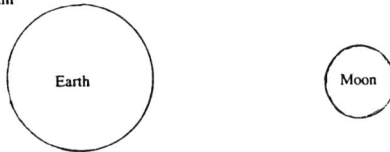

Move the moon closer and this will increase the force of gravity between them.

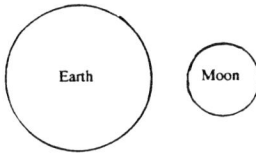

Earth   Moon

Examples of gravity

1 The moons gravity pulling on the oceans on earth to produce tides. If the moon changes its orbit and came closer we could have 100-feet tides.
2 Another example of gravity – in the Arctic and Antarctic gravity causes sheets of ice to be dragged down the mountain to the edge of the land to make glaciers and break off as icebergs.

LIGHT

Light comes from the sun and stars. It is produced by great heat and this is called INCANDESCENCE. It is an electromagnetic radiation which includes radio waves, microwaves, radar, infrared rays, visible light, ultraviolet rays, X-rays and gamma rays. This is called the electromagnetic spectrum. Light travels in a straight line at 186,000 miles per second. Some common properties of electromagnetic waves are:

1 They transfer energy from one area to another area.
2 They can be absorbed by matter and emitted.
3 They do not need a special medium to travel in.
4 They carry no electric charge.
5 They can be reflected and retracted.
6 They can be polarised.

31

Light can be emitted without heat. This is called LUMINESCENCE. eg phosphorescence in the water, fluorescence when certain dyes can absorb ultraviolet light and re-emit it.

Electromagnetic waves

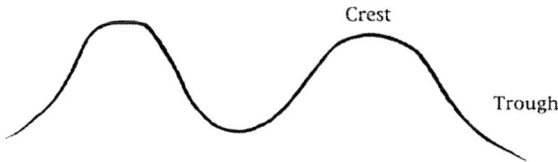

They have crests and trough, like waves in water

Crest

Trough

The distance between the crest is the wavelength in metres

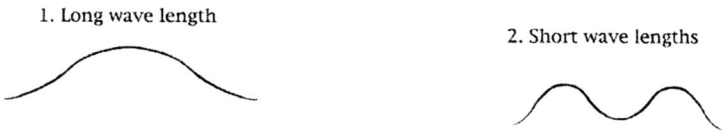

1. Long wave length

2. Short wave lengths

The number of waves per second is the frequency in hertz (Hz).

Radio Waves

This end of the spectrum is divided into bands for different usage eg very low frequency is used for time signals – very high frequency for satellite and space communication. Other frequencies transmit television.

The Visible Spectrum

In chapter one, we have discussed how white light is produced by a combination of many colours. Each of these colours has a different wavelength and make up a rainbow of 7 colours, ranging from red, orange, yellow, green, blue, indigo, violet. The sun shining through raindrops splits the light into these colours. A glass prism does the same thing as the rainbow, splitting white light into 7 colours of the visible spectrum.

Different materials can absorb different wavelengths of light and reflect others. A red object appears red because it reflects red wavelengths and absorbs the other wavelengths.

Mirrors and lenses change the direction in which light rays travel by reflecting the light. A convex mirror gives an image which is larger than the original object, and a concave mirror a smaller image.

The Electric Light Bulb
The hollow glass bulb is filled with an inert gas. Inside is a thin filament of tungsten wire, which heats up and produces light.

Electric Current
This is a flow of electrons through a conductive material eg copper wire.

Direct Current (DC)
This is produced in a battery and the current will flow in one direction only.

Alternating Current (AC)
This current changes its direction of flow at regular intervals.

Domestic Electricity
Supply can convert alternating current to direct current.

EINSTEIN'S GENERAL THEORY OF RELATIVITY (1879-1955)
In 1921 Albert Einstein received the Nobel Prize for Physics for his theory of relativity. Einstein said matter and energy are interlinked. $E=MC^2$. E is the amount of energy produced, M is the mass and C is a constant, the speed of light in a vacuum. Nuclear energy proved Einstein correct. Also Einstein says gravity is connected with space rather than a force of attraction between bodies. He argues bodies make space curve around them and even bends the path of a comet as it passes the sun. Recently experiments have shown the speed of light has changed since the big bang. It has slowed down. This agrees with Einstein's original prediction and constants. The temporal relation between two events can only be analysed by an observer in a state of motion relative to them. I observe with my telescope a happening on Jupiter. Special relativity says information cannot travel faster than the speed of light. Since it takes 35minutes for light to reach my telescope from Jupiter, is the event truly real or a past event? 'The logic of modern physics is to establish the meaning of scientific research and experimentation' (quote from P.W. Bridgemen 1927) The concept of black holes, ultimate

gravity, where time dilation occurs adds another equation into the understanding of space, or is space a concept of our mind?

1927 THE ERNEST SOLVAY INTERNATIONAL PHYSICS MEETING IN BRUSSELS (24–30 OCTOBER)

The meeting was devoted to Quantum Theory. Twenty-nine scientists attended. Nine physicists at the meeting later received the Nobel Prize for their work. The most famous names were:-

Max Planck (1858–1944) – black body radiation
Albert Einstein (1879–1955) – the photoelectric effect
Niels Bohr (1885–1962) – bright light spectra.

When the first engines were built it was noted if heat can be converted to work, it must be a form of energy. In 1847 a German scientist, Hermann Von Helmoltz stated, 'whenever a certain amount of energy disappears in one place, an equivalent amount must appear elsewhere in the same system'. This became the first law of thermodynamics, called the law of conservation of energy. Only three years later another German scientist Rudolf Clausius stated the second law of thermodynamics. Heat flows from a hot body to a colder body and there will be some loss of the total energy which he called entropy. The second law states, the entropy of an isolated system always increases, reaching a maximum at thermal equilibrium (all bodies at the same temperature).

In 400 BC Demucritus described matter as made up of atoms. It was a great idea, but you cannot see atoms, and it remained a talking point until 1820 when new evidence from spectral line emissions suggested atoms might be real.

In the meantime experiments on gases established another important concept. This was the Kinetic theory of gases and an understanding of molecules in motion. This pioneering work by Maxwell in 1860 paved the way for new thinking by Max Planck who was studying radiation and light as waves. Planck said when an object is heated it emits radiation consisting of electromagnetic waves, light, with a broad range of frequencies. If this radiation was confined in a box which absorbed the radiation, and a small hole then allowed the emitted radiation to escape, this would be radiation characteristic of the box. In 1900 this became known as Planck's black body radiation. Planck said matter can absorb and admit electromagnetic radiation, light, in energy bundles called Quanta, whose size is proportional

34

to the frequency of the radiation. His theories were to be discussed at the 1927 conference on Quantum Theory. Einstein entered the arena with Planck. A joint statement said the Quantum relation between the energy of a light photon and its frequency is $E = h \times f$; where h is Planck's constant, a very, very, small amount but not zero, and f = frequency.

Another discovery was made by Niels Bohr working with absorption spectra. Using hot gas Bohr analysed the spectrum, and then using the same gas but cooled. When cool the dark line spectrum has the exact missing bright lines when the gas was hot, ie the cool gas is absorbing the same frequency light. There are energy states in a gas so the gas can take in or give off energy. It was already well known that whereas white light produced the colours of the rainbow, every gas analysed had its own spectral line. This pointed toward the structure of the atom. In 1820 Joseph Fraunhofer studied the solar spectrum and realised the lines were related to the nature of sunlight. Today they are known as the Fraunhofer lines. By 1850 elements in the sun's atmosphere were analysed, including hydrogen and helium.

A Swiss mathematician Johann Balmer discovered a way of organising the findings of spectral lines into an order on the basis of a mathematical prediction which he developed from hydrogen frequencies. It was like an energy diagram because the emission and absorption of light from an atom must correspond to the atom's energy.

Did the scientists have enough knowledge to dissect the atom? Lord Kelvin made a model on the basis the atoms of sun light which behaved according to Maxwell's electromagnetic theory spun around according to Newton's Law of Motion. It was not well accepted. In the Cavendish Laboratories in Cambridge, the atom was being studied by Ernest Rutherford (1871–1937) who was working on radioactive alpha particles. Using a scattering technique where alpha particles were shot at a thin gold foil screen, tiny flashes were produced on a fluorescent screen. Some of these alpha particles were found to bounce back. What collision force could have produced this rebound? It was suggested the atom must have a massive nucleus of positive charge. The new model of the atom showed a nucleus of positive charge and electrons orbiting the nucleus.

In 1912 Niels Bohr joined the Rutherford Manchester Laboratory. Bohr's contributions to Quantum physics was enormous. On page 30 I defined constant angular momentum. Bohr considered this important in constructing his model of the atom, the stationary state of orbits or quantum orbital condition. If an electron is excited it can jump to an orbit where its angular momentum will change. We will give angular momentum the value I and whole numbers to the orbits which we will call the principle quantum number n. Bohr now combined quantum theory E = hf with angular momentum (classical physics). If we know the angular momentum we can calculate the radius and energy of the orbit, using angular momentum quantized in units of h/2rr. Following this Bohr decided to use the energy difference between stationary states to determine the frequencies of the absorption and emission of light using the frequency equation from Planck, Einstein. Blamer had arrived at values for frequencies mathematically starting with the spectra of hydrogen.

Bohr produced a formula using Balmer's data, which when calculated agreed with Bohr's mathematics. This should predict the spectra for all other elements. Like many discoveries all is not accountable.

Arnold Sommerfeld (1868–1951) came to the rescue. He explained the effects of elliptical orbits which occur more often than circular orbits. Another quantum number was introduced K in units of h/2rr. Spectral lines changed again when excited atoms were placed in a magnetic field. Sommerfeld understood the result and added a magnetic quantum number M. This helped but not enough. Something else was happening. To the rescue came a brilliant young theoretical physicist Wolfgang Pauli (1900–1958). Pauli said the electron is spinning, like the earth on its axis. This produced extra angular momentum. So another correction was added, a forth quantum number with two values. Armed with the four corrections Bohr completed his model, stating the chemical and physical properties of an element depend on how the electrons are arranged around the nucleus. Bohr drew up the periodic table using Pauli exclusion principle that each and every electron must have its own set of quantum numbers, and every electron in an atom takes its space in the atom's structure. From hydrogen and helium in the sun come 92 elements on the periodic table. Bohr had completed what is now called 'The Old Quantum Theory'.

In 1924 De Broglie challenged the Quantum thinking. 'The propagation of a wave is associated with the motion of a particle of any sort.' De Broglie's formulae says 'if the wavelength of light is decreased, the momentum of the individual light photon is increased.' Einstein was contacted and said this is a major breakthrough; Bohr agreed.

In 1925 Bohr met a young physicist, Heisenberg, which he admired. Heisenberg's work included mathematically representing corresponding values which was first analysed by. Max Born, an. expert on understanding Matrix Calculus. Another expert on matrix methods Pascual Jordan

36

transposed Heisenberg's theory into a systematic matrix language. This showed that a system of equations could be formulated to produce values of the frequencies and relative intensities of spectral lines. Reisenberg's matrix mechanics was not well received. Erwin Schrodinger was more interested in developing De Broglie's theory of waves, the propagation of a wave is associated with the motion of a particle of any sort, photon, electron, proton etc. The different theories raged on. The great men like Bohr and Einstein were always asked their opinions on different views. Bohr was never completely at home with his own model of the atom and four quantum numbers for corrections.

The day of the Solvay International physics meeting had arrived. It was 24 October 1927. What a collection of physicists and so many theories to discuss. Some agreement was reached. Electrons reveal wave and particle properties. By putting together a collection of ideas there was agreement over the Copenhagen Interpretation (called CHI), which has become the orthodox way of viewing quantum theory.

At the end of chapter one, I have described what might be happening at the edge of the universe. Gravitational forces are very weak or non existent. Dark energy has negative values and these exceed the gravitational field. Where matter reaches this surface it is accelerated into the spacuum (more about this later). Dark energy makes deep inroads into the universe opposing the gravitational field. This causes the universe to fold in on itself, and produces a time difference at a boundary of the fold. If we could identify this boundary it may be possible to cross into a more distant part by a partition called a wormhole into a different time period. It is like viewing multiple universes within a universe. Time will be related to that

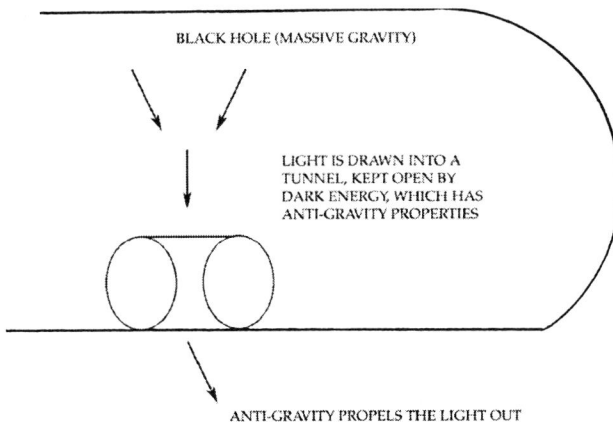

BLACK HOLE (MASSIVE GRAVITY)

LIGHT IS DRAWN INTO A TUNNEL, KEPT OPEN BY DARK ENERGY, WHICH HAS ANTI-GRAVITY PROPERTIES

ANTI-GRAVITY PROPELS THE LIGHT OUT

space. We might be able to observe a past event. How do we create a wormhole? With great difficulty. In the future it might be possible to do it in the laboratory in particle physics. I have drawn a picture of a fold in the universe. For light to travel round the fold would be a great distance, to travel across it, a short distance.

Time is related to space. Distance is related to space. Time and distance are tied up with relativity. For convenience we invent a time period.

For example; $\dfrac{\text{Distance to go}}{\text{Speed made good}}$ $\dfrac{20 \text{ miles}}{10 \text{ mph}}$ = two hours of time

I defined a spacuum as a space not measured by time and distance. It might be measured by a difference in time X a relativity constant (which we don't know).

We can look back in time at past events. The light from distant objects has taken so long to reach us, the objects may have ceased to exist long ago. To travel through a wormhole tunnel into the past, would be the present in that space and time, and you could not influence the past. If you could travel through a wormhole tunnel into a more advanced civilisation and return, you could come back armed with great knowledge.

What Do We Know About the Time Clock?
Time is a spacing and different means are necessary to measure this spacing. Our sun gives us a time clock by which we measure day and night. We can observe an object, for example a tree. This is an event. Now we notice the tree has a shadow from the sun, and after some hours the shadow has moved. An event in time has occurred and the spacing of the shadow is a measure of the sun's movement. The first clocks were sun dials to record this spacing. In 24 hours the sun maps out 360° which is 15° per hour. It was necessary to develop accurate clocks to help ship's navigators to find their longitude, west or east of the Greenwich Meridian.

Time in nature has spacing for growth, reproduction, migration, hibernation, often related to the seasons of the year, which again are related to temperature, water and food supplies.

The body has several clocks for the 24 hour 'Rhythm of Life', spacing for eating, working, sleeping, etc. Other clocks are designed for specific purposes eg the time judgement for a high jumper to take off and leap into the air. This time clock is located in part of the brain called the hypothalamus.

We watch our children grow up and record the event on film. Years go by and we show the film – that's when you were very young, now you are in your teens and leaving school. This is your first job, and look now, you are getting married; so on to middle life and old age. We associate these time intervals with our life span of three score years and sixteen. We can run the film backwards and say that's in the past, but we cannot grow younger.

Musicians can develop a great sense of time. This suggests we can develop timing for different purposes. Could we learn more about time by developing our senses?

We have developed clocks to measure very small intervals of time, called atomic clocks. Einstein predicted in his general theory of relativity that time is relative to speed and gravity. We have other times besides earth time. If we leave planet earth in a space craft and accelerate to three quarters of the speed of light (if that is possible), returning after five years, there will be a difference in the time between persons in the space craft and those on earth. Earth could have aged hundreds of years. Time is slowed by massive gravity. If you could orbit your spacecraft near a black hole without falling in, time would nearly stand still relative to earth. If you could return to earth, earth may have aged 100,000 years!

If the earth is an experimental laboratory, what a convenient experiment to return every five years to see the earth that may have aged 100 years or more.

If the original creator is still with us after 16 billion years, he is not likely to be made of flesh and blood. He is more likely to be a 'super something else' connected to a big computer. Certainly our creator didn't want to blow himself up with the big bang, and is likely to have distanced himself at a safe distance.

When we build a house we call in many experts in different fields, from bricklayers, electricians, plumbers etc. When the creator designed every-thing, did he have a team of advisers, and did they all agree with each other.

The idea of a creator having assistants is in keeping with the plurals for the gods in Genesis! Such as chapter 3 V 5... 'ye shall be as gods.' And v 22 'behold the man has become as one of us, to know good and evil ......' If we are to believe in the Old Testament, the creator had many assistants and not all good.

The qualification of a Nobel Prize in Quantum Physics, must surely be a passport to heaven.

Further books I recommend

1 Science Facts by Steve Setford – Pockets, Dorling, Kindersley, London WC2E BPS
2 Dictionary of Nature by David Bumie – Dorling, Kindersley, London WC2E BPS
3 The Usborne Living World Encyclopaedia by Leslie Colvin and Emma Speare – Usborne Publishing, London ECIN BRT
4 The Fires Within – Volcanoes on Earth and Other Planets by John Murray – Dragons World Ltd, Limpsfield, Surrey RHS UDY
5 Quantum Theory by J P Mcevoy and Oscar Zarate – Kun Books UK and Totem Books USA.

Chapter 3

Meteorology
How the weather is formed

Introduction

The weather is different on different parts of the earth's surface, colder near the polar regions, warmer near the equator. Also it changes with the seasons of the year. Our summer is Australia's winter. We can either adapt to these changes or do as some animals and birds do, migrate to a more suitable climate especially if there is going to be a lack of rainfall. The unequal heating of the surface of the earth makes short term and long term changes, periods of cold and warm which change over thousands of years. We are in a period of global warming but 10,000 years ago, at the end of the last ice age, Europe was frozen, and polar bears roamed the South of France. Meteorologists haven't kept records long enough to be able to predict these long term trends, but there is concern that man may be contributing to the heating of the earth by 'greenhouse' gases e.g. car fumes, industrial fumes, polluting the atmosphere.

What are the heating and cooling processes which affect our weather, and the earth's temperature? 1) the sun, 2) volcanoes, 3) the temperature of the oceans, 4) the atmosphere, wind, rain and rotational forces.

Meteorology is an excellent example of all the forces of nature at work. When too many variables come into the equation a state of chaos reigns only to return to a predictable pattern when forecasts can be made with a reasonable degree of accuracy. When you look at weather forecasts on TV, hopefully you will be able to understand what is happening after reading this chapter.

Meteorology

The sun heats up the earth's surface unequally depending on the seasons of the year, the latitude of the observer, and whether it heats land or sea.

41

Temperature varies with:-
1 Latitude
2 Seasons of the year
3 Height
4 Prevailing wind
5 Amount of cloud present
6 Nature of the surface.

The difference in temperature in different parts of the earth gives rise to a difference in pressure, and the wind is the movement of the air across the surface of the earth between pressure areas, tending to blow from high to low; the steeper the pressure gradient the stronger is the wind.

PRESSURE SYSTEMS AND WIND DIRECTION

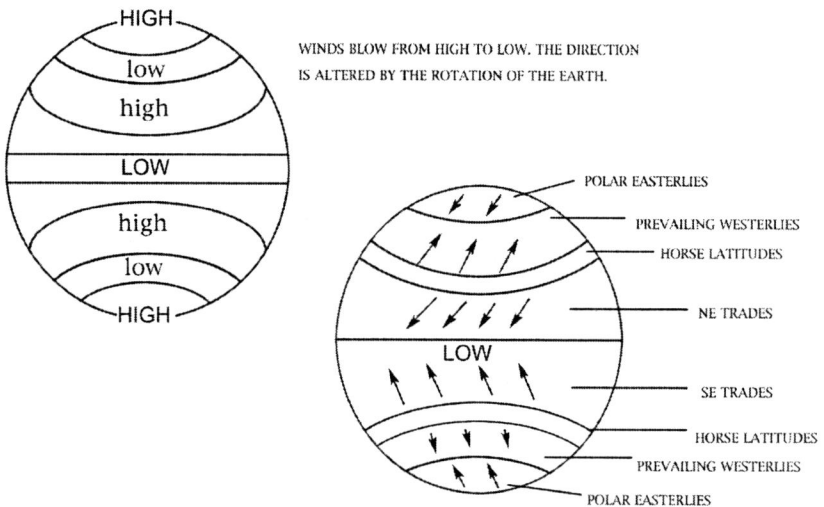

WINDS BLOW FROM HIGH TO LOW. THE DIRECTION IS ALTERED BY THE ROTATION OF THE EARTH.

HIGH
low
high
LOW
high
low
HIGH

POLAR EASTERLIES
PREVAILING WESTERLIES
HORSE LATITUDES
NE TRADES
LOW
SE TRADES
HORSE LATITUDES
PREVAILING WESTERLIES
POLAR EASTERLIES

The sun causes water to be evaporated from the oceans, lakes and rivers. The water vapour in the atmosphere is a balance between evaporation and precipitation (rain). Air masses of different temperature collide and don't mix easily like running your bath with hot and cold water, it needs stirring, cold air undercuts warm air and sets up turbulence called depressions. Associated with these low pressure areas are rising air currents, rain, strong winds, unstable air. Other areas of plentiful sunshine form areas of high

42

pressure e.g. The Azores, Associated with high pressure is air falling, fine weather, lighter winds, stable air. The rotation of the earth causes the air to move from west to east. These geostrophic forces cause air to be deflected to the right in the northern hemisphere, and to the left in the southern hemisphere. The bath water, as it runs out, spins to the right in north latitudes, and spins to the left in the southern hemisphere.

Wind blows out of the Equator and into the Poles. Winds nearer the surface blow from an area of high pressure to one of lower pressure, i.e. from the poles toward the equator being deflected by the earth's rotation.

The low pressure over the equator is called the intertropical convergence zone = ITCZ It follows the sun, moving to latitude 12° north in summer and south of the equator in winter, associated with it are cumulonimbus clouds towering to 65,000 feet.

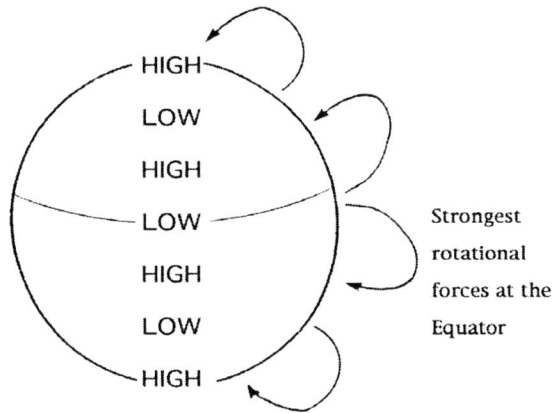

HIGH
LOW
HIGH
LOW — Strongest
HIGH — rotational
LOW — forces at the
HIGH — Equator

The air up to 5 miles is called the troposphere and contains all the weather of clouds, wind, rain etc. There is a decrease in temperature of air in height up to 5 miles.

The Seasons of the Year
The sun moves from Declination 21 N – Declination 21 S. June it is summer in England. June it is winter in Australia.

Nature of the Surface Being Heated
The soil is a poor conductor and heating takes place near the surface. Much more heat is required to raise the surface of the sea.

Amount of Cloud Coverage
Cloud coverage can prevent heat loss to space. Temperatures over the land are higher by day than over the sea. At night however the land loses its heat especially if there is no cloud coverage, but the sea retains the heat.

Sources of Air Masses

The direction from which the wind is blowing will give an idea as to the nature of its moisture content and temperature e.g. a south wind blowing all the way from the Sahara Desert having travelled across the continent of Europe will be a dry warm wind This will be called a continental wind and travelling from the Tropics, Tropical Continental – TC. A south west wind will have travelled across the sea and be called maritime – from the Tropics, Tropical Maritime – TM. This will be a warm moisture laden wind. A north wind blowing over Scotland in winter from the Arctic will be called Arctic Maritime – AM. This is a very cold wind, moisture laden, which on hitting the high mountains of Scotland will reach saturation point and precipitate as snow. However by the time this north wind reaches the Isle of Wight it will have lost most of its snow. A polar continental wind is a bitter cold dry wind from Siberia.

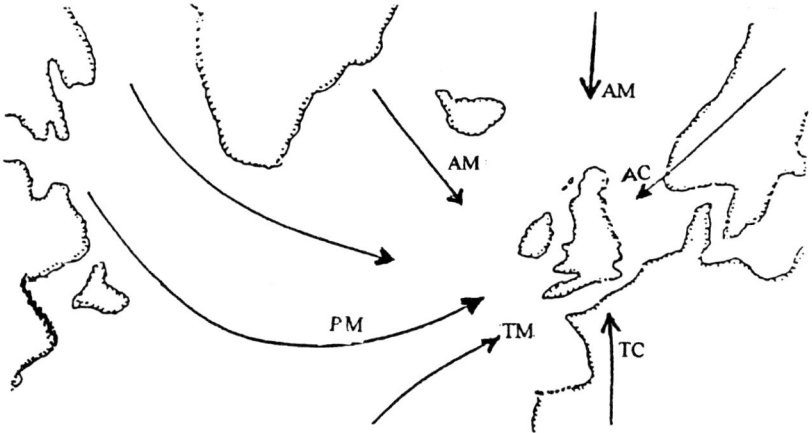

TM = Tropical Maritime   AC = Arctic Continental
TC = Tropical Continental   PC = Polar Continental
AM = Arctic Maritime

Below is a list of definitions and subjects which are involved in the study of meteorology.

Definitions

1   Atmospheric Pressure
2   Millibars
3   Isobars

4 Properties of the Atmosphere
5 Prevailing Wind, Temporary Winds
6 Wind Speed, Beaufort Scale
7 Sources of Air Masses
8 Rotation of Earth
9 Fronts – Cold – Warm Occluded
10 Depression
11 Characteristics of an Approaching Depression
12 Weather Associated with Fronts
13 Buys Ballot Law – Lee Shore – Weather Shore
14 Theory of Cloud Formation) Cloud Types for Observers, HMSO
15 Types of Clouds and Heights) Publications Centre, P0 Box 276 London SWS 5DT
16 Rain due to (1) Convection, (2) Turbulence, (3) Orographic lifting, (4) Frontal
17 Anti Cyclone
18 Land Sea Breeze Effects
19 The Effect Gradient Wind has on the Formation of Sea Breeze
20 Types of Fog
21 Sources of Publications
22 Principle of Forecasting, Isobars and the Wind Pressure Relationship, the Barometer
23 Forecasting at Sea and drawing a Synoptic Chart
24 Wind Direction in Strong Gusts. The effect the land has on the wind and rain.
25 Katabatic, Anabatic Winds.
26 Hurricanes, Typhoon, Monsoon, Water Spout, Tornado
27 Special winds – e.g. Mediterranean (1) Mistral, (2) Sirocco, (3) Bora, (4) Fohn, etc.
28 Special Characteristics – (1) Secondary Lows, (2) V Depressions, (3) Col.
29 Wind Rose on Routing Charts
30 Climatological Normals or Averages
31 Ocean Passages and Semi-permanent Wind Directions
32 Charts of Marine Meteorology Sea Surface Currents and Ice
33 Researches and Applied Meteorology
34 Meteorological Instruments – Barometer, Hydrometer, etc.
35 El Nino and La Nina
36 Global Warming.

A new meteorological building in Exeter, in England, has three massive computers to analyse data. The new orbiting weather satellite is sending back much detailed information. We have geostationary satellites, observing one area continuously. This is the beginning of the new age of meteorology and more accurate forecasting.

Sources of Forecasts
1 Radio Forecast – land forecast or a sea shipping forecast.
2 There may be a forecast in the newspapers.
3 There may be a television forecast for your area.
4 There may be a fax print out of a Synoptic Chart.
5 The local airport may issue a weather report.
6 If you are by the sea, the coastguard may have details.
7 Pre-recorded weather forecasts on the telephone.
8 The internet.
9 Special weather forecast transmitting stations.

1 Atmospheric Pressure
The force exerted by the atmosphere at any point on the earth's surface. It is measured by a barometer and converted to sea level

2 Millibars
Atmospheric pressure is measured in Millibars and decreases with height.

3 Isobars
Lines drawn on weather maps through places having the same mean sea level pressure. They are drawn at 4 Mb intervals i.e. 1004 Mb, 1008 Mb, 1012 Mb. Isobars never cross each other and are frequently found to form closed curves around an area of high or low pressure.

Gradient Wind
If there was no rotation of the earth, the pressure gradient wind would blow directly from a place of high pressure to low pressure at a speed dependent on the difference of pressure between the two places i.e. like an ordinary gradient, objects roll downhill.

Geostrophic force (Coriolis Effect)
This deflects the gradient wind to the right in the Northern Hemisphere. The earth is a sphere rotating eastwards on its axis. Different parts of the

earth's surface have different actual eastward speeds, depending on the latitude, and lower latitudes have higher speeds than the higher latitudes. The speed at the equator is a maximum and the speed at the two poles zero.

The geostrophic force is opposed and counter-balanced by the gradient wind force. Both of these forces are at 90 degrees to the wind direction.

4 Properties of the Atmosphere

(1) Temperature, (2) Pressure, (3) Wind, (4) Water Vapour. Temperature has the direct influence on forming weather. The unequal heating of the earth gives rise to different pressure areas. The wind blows across a pressure gradient. The amount of water vapour that can be carried by an air stream depends on (a) how dry is the air, (b) what temperature is the air, (c) how quickly will saturation take place and the dew point be reached.

There is a decrease in temperature with height, the rate of decrease is the lapse rate (unless an inversion occurs). Changes in temperature affect the atmospheric pressure which effects the direction and strength of the wind. Air can contain a varying degree of moisture content and is constantly changing, e.g. evaporation from the surface of the sea. Saturation of the air takes place at dew point.

Unstable Air	*Stable Air*
Cumulonimbus	Stratus-type cloud
Cumulus type clouds	Steady winds unless very strong
Gusty winds. Bumpy air	'Smooth' air
Good visibility	Moderate or poor visibility

5 Wind – in the Northern Hemisphere

The rotation of the earth from west to east drags the earth's atmosphere with it. The very high winds called the jet stream circle the earth from west to east. The clockwise circulation in the North Atlantic and Pacific cause predominantly westerly winds to flow at latitude 40°N. At latitude 18°N the return airflow to the Caribbean causes easterly winds called trade winds.

Wind moves anticlockwise around lows and clockwise around highs.

eg

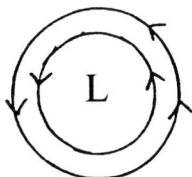

In reverse in the southern hemisphere.

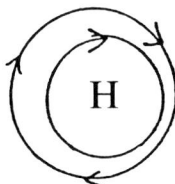

47

Permanent winds – e.g. Trade winds blow most of the time in the same direction, producing surface drift currents.

6 Wind Speed

The Beaufort Scale goes from 0–12.0 is a calm, 12 a hurricane. An average sailing breeze when all sail can be carried is Force 4 which is 11–16 knots and the open sea has only small waves. Force 6, 22–27 knots produces large waves about 10 ft with some breaking crests coming aboard. Going to windward this will be uncomfortable, and reefing sails are necessary. Force 8 is called a gale – average wave height 23 ft. At sea, ships are damaged by the force of large breaking waves, which in hurricane wind conditions, when wind speeds above 100 knots are likely, can be 80 feet high.

7 Sources of Air Masses

This has been described on page 44.

8 Rotation of the Earth

This causes the geostrophic force, and air is deflected to the right in the Northern Hemisphere and to the left in the southern hemisphere. Friction at the surface causes air to be inclined at an angle to the isobars, inwards on lows, outwards on highs. This is called the angle of indraft, 10°–20° over sea, 30° over the land.

9 Fronts

A front is the intersection of a frontal zone with the surface, the frontal zone being the boundary between two adjacent air masses. Temperature is used to distinguish fronts.

Cold Front – cold air replacing warm air at the surface.
Warm Front – warm air replacing cold air at the surface.

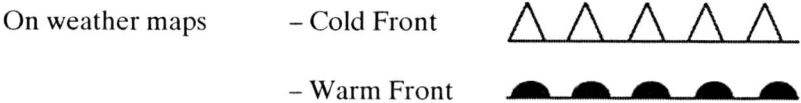

On weather maps   – Cold Front   △ △ △ △ △

                    – Warm Front   ● ● ● ● ●

Eventually the warm air and cold air will mix and form an occluded front.

△●△●△●△●△

6 Wind Speed, Beaufort Scale

Beaufort Number	Mean Velocity Knots	m/s	Descriptive Term	Deep Sea Criterion	Probable height of waves in feet
0	Less than 1	0-0.2	Calm	Sea like a mirror	
1	1-3	0.3-1.5	Light air	Ripples with the appearance of scales are formed but without foam crests.	
2	4-6	1.6-3.3	Light breeze	Large wavelets, still short but more pronounced. Crests have a glassy appearance and do not break.	½
3	7-10	3.4-5.4	Gentle breeze	Large wavelets. Crests begin to break. Foam of glassy appearance. Perhaps scattered with white horses.	2
4	11-16	5.5-7.9	Mod. breeze	Small waves, becoming longer: fairly frequent horses.	3½
5	17-21	8.0-10.7	Fresh breeze	Moderate waves, taking a more pronounced long form; many white horses are formed. (Chance of some spray).	6
6	22-27	10.8-13.8	Strong breeze	Large waves begin to form; the white foam crests are more extensive everywhere. (Probably some spray).	10-13
7	28-33	13.9-17.1	Near gale	Sea heaps up and white foam from breaking waves begins to be blown in streaks along the direction of the wind.	18
8	34-40	17.2-20.7	Gale	Moderately high waves of greater length; edges of crests begin to break into spindrift. The foam is blown in well marked streaks along the direction of the wind.	23
9	41-47	20.8-24.4	Strong gale	High waves. Dense streaks of foam along the direction of the wind. Crests of waves begin to topple, tumble and roll over. Spray may affect visibility.	25
10	48-55	24.5-28.4	Storm	Very high waves with long overhanging crests. The resulting foam in great patches is blown in dense white streaks along the direction of the wind. On the whole the surface of the sea takes a white appearance. The tumbling of the sea becomes heavy and shock-like. Visibility affected.	29
11	56-63	28.5-32.6	Violent Storm	Exceptionally high waves. (Small and medium-sized ships might be for a time lost to view behind the waves). The sea is completely covered with long white patches of foam lying along the direction of the wind. Everywhere the edges of the wave crests are blown into froth. Visibility affected.	37
12	64 +	32.7 +	Hurricane	The air is filled with foam and spray. Sea completely white with driving spray; visibility very serious.	45

BEAUFORT TABLE

49

10 Formation of a Front and Depression in Northern Hemisphere

⟶ Cold Polar Air

A — — — — — — — — — — C   AC = surface boundary of a front

⟶ Warm Tropical Air

A Small wave depression forms at B, cold air undercutting warm air.

1

B
A   C

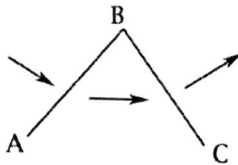

Circulation around B takes place

2

B
A   C

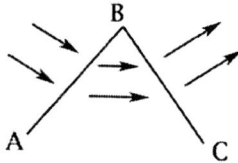

AB = Cold front
BC = Warm front

3   Low Forms

B
cold air   cold air   Y
X
warm air

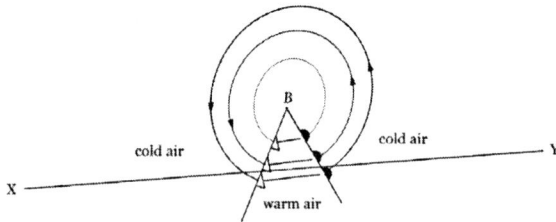

CuNb = Cumulonimbus   Ci = Cirrus   Cs = Cirro-stratus   As = Altostratus   Ns = Nimbostratus

50

Section Through XY

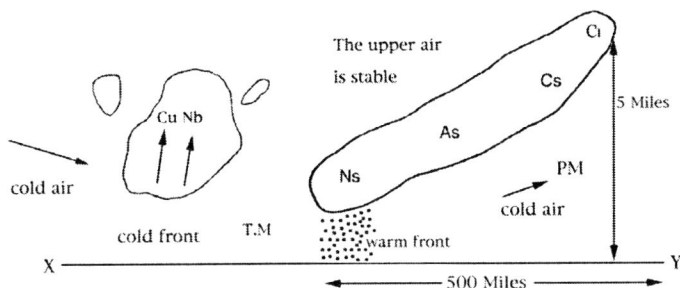

An average low will travel at about 25 knots and an observer at Y will be about 500 miles from the low. It will take 24–36 hours for the low to move.

The first signs at Y are high cirrus clouds about 18,000 feet, thin filaments resembling mares' tails. (Speed of cirrus – clouds approaching not necessarily a good guide of severity of depression). Barometric pressure falls. Cirrostratus clouds cover the sky, a whitish transparent cloud, giving a halo round the sun and moon. Cold air is moving ahead of the warm front so the temperature falls. The low develops and deepens. The cloud base becomes lower. Altostratus forms – a greyish or bluish cloud covering the sky between 8,000–18,000 feet. The barometer falls. Scudding black clouds form called Nimbostratus, low clouds up to 8,000 feet, and it starts to rain. The warm front has arrived. Visibility deteriorates in the rain, the temperature starts to rise, and humidity increases. The upper air is stable. The wind veers SW. The barometer continues to fall. As the warm front passes the rain stops, the skies clear, the barometer rises, and the wind veers to NW with the likelihood of an increase in velocity. The temperature falls as cold air arrives. Towering Cumulonimbus clouds form rising to 20,000 feet causing violent updrafts of wind and unstable air. There is a rapid fall in humidity, a great improvement in visibility except in the rain and hailstorm. The wind will slowly decrease and showery weather with a westerly wind is usual. The barometer will stay steady or rising slowly until the next low moves in or high pressure forms. Often a backing of the wind to SW and a falling barometer will indicate another low arriving especially in winter time.

Cold Front Weather
Depressions tend to move from West to East and track NE. Eventually the warm and cold air will mix and the depression will fill and disappear. The

51

cold air 'catches up' the warm air and an occlusion forms. A cold front may pass a long way south of the original low. In winter cold fronts pass south of the Canary Islands and travel at about II knots from West to East. Depressions never pass south of Latitude 20°N.

In the area off Bermuda cold air is seen undercutting the warm trade wind clouds which climb over the top of them. Cold front weather is characterised by Cumulonimbus clouds (thunder clouds).

An Occlusion

The type of weather associated with an occlusion varies and may be of the warm front or cold front type.

(1) Warmer air at the back

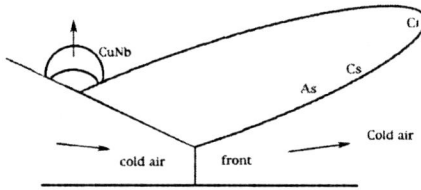

(2) Colder air at the back.   (3) Trough of low pressure.

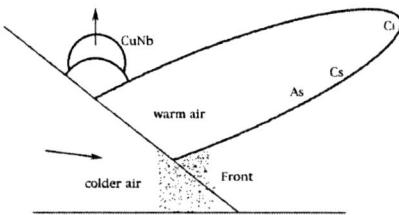

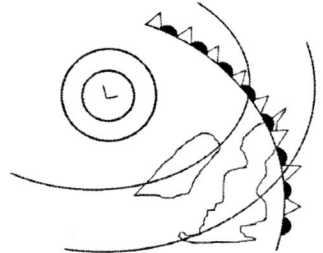

Buys Ballot Law

Put your back to the wind and the depression is on your left hand side (in the Northern hemisphere). The winds are going anticlockwise around a low and converging toward the centre. If the low passes north of your position the SW wind will be replaced by NW winds.

Leeshores (For Sailors)

If you anchor and the wind changes direction it could blow you up the beach onto a leeshore. The anchorage is no longer sheltered and you must move away. It is very important to know if the depression is passing north or south of you. An example is given for a depression passing North and South of your anchorage in Studland Bay, off Poole Harbour, on the South Coast of England. This is also important for campers, holiday-makers, to choose a sheltered spot in case the depression passes south of you and brings wind in from the south east, with a risk of flash flooding. If a depression passes close to you the change in wind direction of 90° can cause big seas developing at right angles to the existing swell.

An example of this is the Fastnet disaster in 1979
This is a yachting race from Cowes on the Isle of Wight, westward, rounding Lands End, and then north to round the Fastnet Rock off Ireland before returning to finish at Plymouth. In 1979 extreme bad weather conditions caused 15 people to die on the race. An area of low pressure, fuelled by cold air at the termination of the jet stream, arrived off Fastnet Rock. The gale force SW winds of 40 knots were replaced by NW winds as the low passed by: The tightly packed isobars produced winds of 70 knots. The 30 feet Atlantic swells from the SW were met by seas from the NW at right angles. The breaking high waves made it impossible for some yachts to cope. Yachts were driven down into the sea, dismasted and rolled over. The Irish Sea is inside the continental shelf with depths of 200 feet in places. Surface

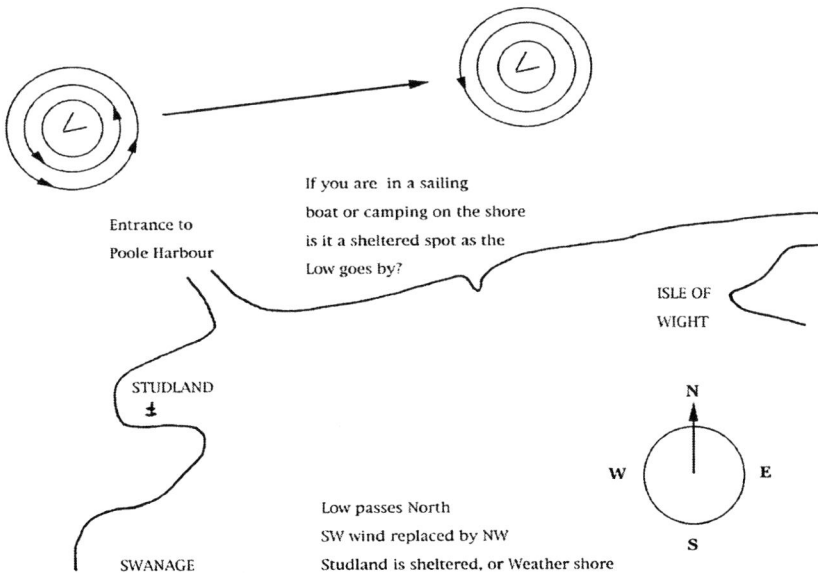

Entrance to
Poole Harbour

If you are in a sailing boat or camping on the shore is it a sheltered spot as the Low goes by?

ISLE OF
WIGHT

STUDLAND

Low passes North
SW wind replaced by NW
Studland is sheltered, or Weather shore

SWANAGE

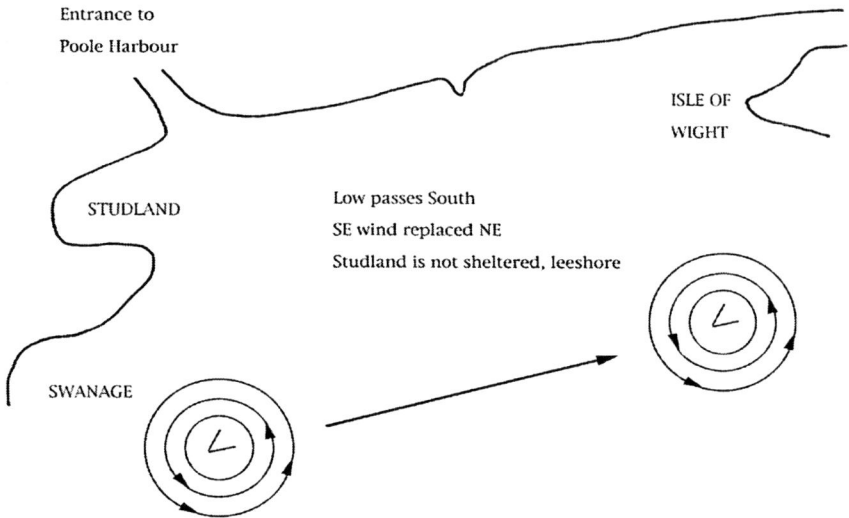

Entrance to
Poole Harbour

ISLE OF
WIGHT

STUDLAND

Low passes South
SE wind replaced NE
Studland is not sheltered, leeshore

SWANAGE

friction slows up the waves and makes them steeper. The shipping forecast at 5.50 pm indicated bad weather. A further bulletin at 6.20 pm indicated a worsening situation with force 10 winds. Those yachts which heard this bulletin had the chance to seek shelter in Falmouth Harbour.

The Theory of Cloud Formation

When moist air rises it cools. It may become saturated and on further cooling condense with the formation of cloud.

Clouds will form at different heights. The formation of cloud doesn't mean it will rain. Rain will occur if the droplets in the cloud overcome the frictional resistance of the air and fall to the ground. Within a cloud there may be snow, ice, supercooled droplets, water droplets. High clouds like Cirrus clouds are mostly ice crystals.

There are two theories on precipitation from clouds:-

1. Bergeron – water droplets fall out of the cloud as rain from snow and ice and supercooled droplets above.

54

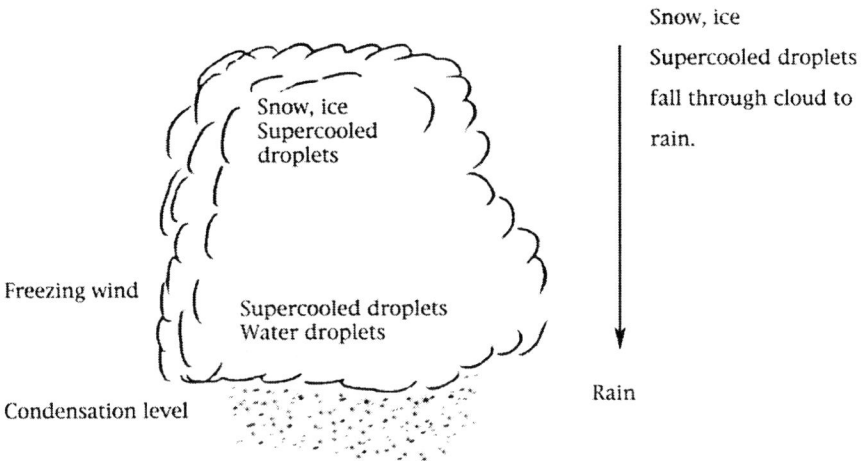

Snow, ice
Supercooled droplets
fall through cloud to
rain.

Rain

2. Coalescence Theory

Water droplets vary in size and large ones attract smaller ones until eventually they fall to the ground as rain, i.e. it can rain from a cloud without supercooled droplets and ice forming.

In the International Cloud Atlas clouds are classified:-

Types of Cloud and Their Heights
1 Genera   – 10 main types.
2 Species   – 14 main types
3 Varieties   – 9 main types
4 Supplementary features (6) and accessory clouds (3).

Clouds can also be classified into heights.

1 Genera – 10 Types

(1) High Clouds	(2) Middle Clouds	(3) Low Clouds
(above 18,000 ft)	(8,000–18,000 ft)	(Below 8,000 ft)
Cirrus Alto Stratus	Stratus Cumulus	
Cirrocumulus	Altocumulus	Stratus
Cirrostratus	Nimbostratus	Cumulonimbus

Low to very high clouds – marked vertical ascent = Cumulonimbus. The types of clouds present may indicate the type of weather expected. 'Cloud types for observers' is recommended.

Rain

1 Convection rain – moist air carried upwards in Cumulonimbus clouds – heavy rain and thunderstorms.

2 Turbulence rain – light rain or drizzle from turbulence clouds.

3 Orographic – high ground, air rises and precipitates, so western parts of Britain wet, eastern parts drier. Stratus cloud forming on the windward side of high ground rain can last for a long period of time.

4 Frontal rain – passage of warm front.

Anticyclones – average pressure summertime British Isles 1024 Mb. Areas of high pressure, winds circulating clockwise around highs, weak pressure gradient, light winds, clear skies (unless high is over sea and has picked up Stratus cloud), Fine weather.

Highs (1) Semi-permanent e.g.) Azores.
　　　(2) Temporary.

Movement of air in a high

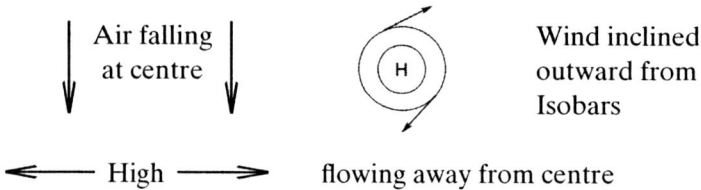

| Air falling at centre | H | Wind inclined outward from Isobars |

←— High —→　flowing away from centre

The characteristic of the weather will depend on the position of the high.

High over N France will be bringing warm south winds from the Sahara Desert, very dry, clear skies, light breezes. High over N Sea will bring moisture, low Stratus cloud.

The Azores High

In the third week of May the Azores high intensifies and expands 500 miles west towards Bermuda. The clockwise rotating winds send a ridge of high pressure towards Western Scotland recurving with high pressure towards the north coast of Spain.

The high over the North Pole drifts down towards Spitzburgen and high pressure builds over the African coast and Europe. If all these highs meet up together and intensify you will have a special fine summer as in 1976. The easterly winds produced reached America as high as latitude 43°N.

56

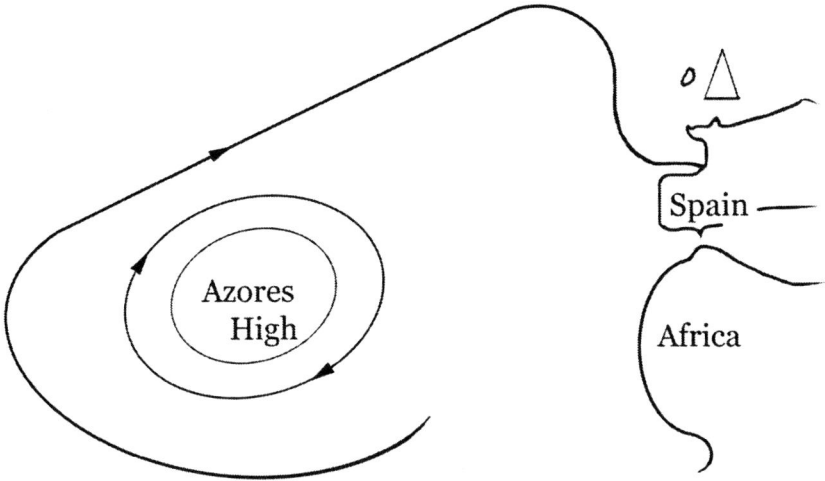

The wind rose on routing charts give the average wind values for each month.

WIND ROSES are shown in red. The arrows fly with the wind their length indicates percentage frequency on the scale –

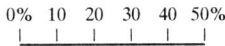

their thickness indicates force

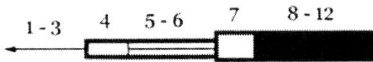

The frequency scale is 2 inches to 100%. From the head of the arrow to the circle is 5% and provides a ready means of estimating the percentage frequency. The number of observations is shown by the upper figure, the percentage frequency of variable winds by the middle figure, and calms by the lower figure in each rose.

MONTH OF AUGUST   MEAN SEA TEMPERATURE

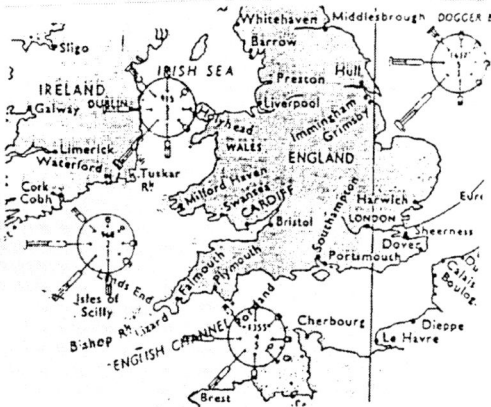

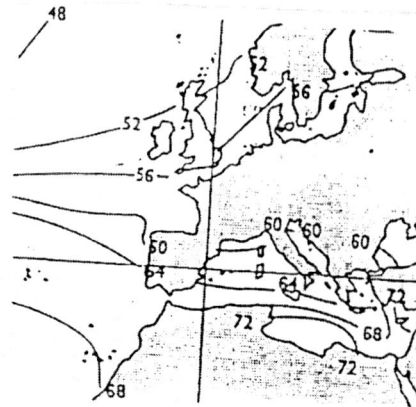

The Land Sea Breeze Effect

Sea breeze with no gradient wind
1   Air over land is warmed and expands – rises (Max 600m)

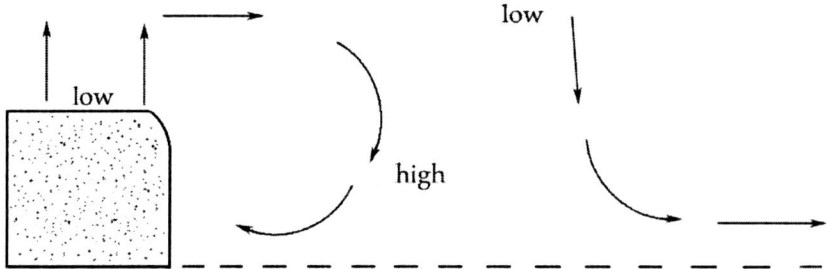

2   Air drifts seawards
    ¼– ½ mile.
3   Air falls and returns inshore (Causes dissolving of low cloud just
offshore).
4   Breeze increases and extends seawards (Force 2).
5   Direction of breeze turns right and by mid afternoon the direction is
    nearly along the coast. A shift of 50–60 degrees; a slight breeze may
    drift seawards for many miles.

6 Sea breeze dies away towards sunset, depending on the rate of fall of temperature over the land.

With no gradient wind the sea breeze will be light and sail ½ mile from the shore to get maximum breeze.

BAYS   ISLANDS

Less Thermal Activity

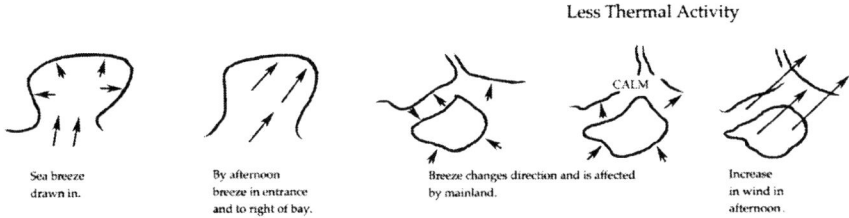

Sea breeze   By afternoon   Breeze changes direction and is affected   Increase
drawn in.   breeze in entrance   by mainland.   in wind in
                  and to right of bay.   afternoon.

Barometer

The onset of a gale is often heralded by a fall of the barometer and by a backing of the wind, but sometimes it is associated with a rapidly rising barometer and veering wind. In the British Isles a rapid rise or fall only brings force 8 in one in three occasions. Gales sometimes come without much fall of the barometer. A moderate easterly wind with a slight fall of barometer should be looked on with suspicion. Southerly gales are most likely with a falling barometer and north westerly gales with a rising barometer. If the pressure is above 1020 Mb and steady or rising quiet weather is likely for 24 hours. In summer it lasts longer, in winter shorter. The wind is usually strongest in the warm sector and decreases on passage of the cold front. On occasions the wind rises to a maximum shortly before or actually as the cold front passes, and continues for several hours to blow harder than it did in the warm sector.

Barometer Indications in General (Not below latitude 20° N of the Equator)

1 *Low Pressure* shows unstable and changing conditions.
2 *High Pressure* shows stable and continuing good conditions.
3 *Steady Rise* shows cold weather approaching.
4 *Steady Fall* shows bad weather approaching.
5 *Rapid Rise* shows better weather may not last.
6 *Rapid Fall* shows stormy weather approaching rapidly.

Pressure Changes

Slowly – less than 1.5 Mb in 3 hours.
Quickly – 3.5–6 Mb in 3 hours
Very rapidly – more than 6 Mb in 3 hours.

Single Observer Forecast at Sea

1 Watch barometer every hour – change of 8 Mb in 3 hours – gale.
2 Change in direction and strength of wind – Buys Ballot Law.
3 Change in appearance of sky – cloud type and amount of sky covered – Okras.
4 Humidity – wet and dry bulb thermometer – surface sea temperature giving alterations in visibility.
5 Change in temperature (air temperature useful in recognising type of air stream).
6 Seamen's Rhymes. a) Red sky at night shepherds delight. Red sky in morning shepherds warning.
   b) Mackerel sky and mares tails cause lofty ships to carry low sails.

Principles of Forecasting

Take Down the Forecast.

You may need to write it down: I have enclosed a form used by seamen around the British Isles showing sea areas defined and coastal report stations. (Courtesy of the Royal Yachting Association and Royal Meteorological Society).

1 Any gale warnings, strong wind warnings.
2 The general synopsis, the time of the general synopsis, the position of lows, highs, fronts, barometric pressure of highs and lows, and expected movements and changes in centre of pressure. Also write down the forecast wind direction, wind speed, visibility and barometer readings.
3 The coastal reports are usually four hours after the general synopsis which will need updating. They give barometric pressure readings which will assist in constructing the synoptic chart and joining lines of equal barometric pressure. If one place says SW winds, rain, poor visibility, falling barometer, you are in a warm sector. If another place says NW winds, gusty, visibility good except in showers, barometer rising, you are in a cold sector.

RYA / R Met Soc Metmap (Courtesy of the RYA and R. Met Society)

GENERAL SYNOPSIS .... at .... GMT/BST ....

Gales	SEA AREA FORECAST	Wind	Weather	Visibility
	Viking			
	N. Utsire			
	S. Utsire			
	Forties			
	Cromarty			
	Forth			
	Tyne			
	Dogger			
	Fisher			
	German Bight			
	Humber			
	Thames			
	Dover			
	Wight			
	Portland			
	Plymouth			
	Biscay			
	Trafalgar			
	Finisterre			
	Sole			
	Lundy			
	Fastnet			
	Irish Sea			
	Shannon			
	Rockall			
	Malin			
	Hebrides			
	Bailey			
	Fair Isle			
	Faeroes			
	SE Iceland			

COASTAL REPORTS (Shipping Bulletin) at .... BST GMT	Wind Direction	Force	Weather	Visibility	Pressure	Trend
Tiree						
Stornaway						
Sumburgh						
Fife Ness						
Bridlington						
Dover						
Greenwich LV Auto						
Jersey						
Channel LV Auto						
Scilly Auto						
Valentia						
Ronaldsway						
Malin Head						

11/95

COASTAL REPORTS (Inshore Waters) at .... BST/GMT				
Boulmer				
Bridlington				
Walton on the Naze				
St Catherine's Point				
Scilly Auto				
Mumbles				
Valley				
Liverpool Crosby				
Ronaldsway				
Killough				
Larne				
Machrihanish				
Greenock				
Benbecula Auto				
Stornoway				
Lerwick				
Wick Auto				
Aberdeen				
Leuchars				

Further Plotting Details – Procedure on weather map provided for a sea forecast

1 Plot any lows, highs, fronts advanced as indicated.
2 Plot wind directions for each sea area. This is a line in the direction from which the wind is blowing in towards the station circle.

SW 5 – 1 feather = 2 Beaufort force
½ feather = 1 Beaufort force

3 Plot wind speed and direction for next 24 hours in each sea area.
4 Draw the isobars for the pressures at coastal stations. Join places of equal pressure. When you have a pattern, join up the broken sections. Isobars are drawn slightly inwards on lows and outwards on highs. So the wind direction on each sea area helps you to construct the lines. Try to complete your weather map at 4 Mb intervals. Use the geostrophic scale on the weather map to help you.
5 Remember the closer the isobar spacing the stronger the wind. Wind blows across a pressure gradient. You could have several lows close to each other with little pressure gradient between them.
6 You must recognise from the change in wind direction where the cold fronts are positioned. e.g. If one sea area shows SW wind and the next NW winds the cold front is between the two areas.

Taking Down the Shipping Forecast from the Radio. Example given (extracts)

The shipping forecast issued by the Meteorological Office at 2330 on Saturday 31 April 1999.

There are warnings of gales in Viking, North and South Utsire, Forties, Cromarty, Forth, Tyne, Fisher, German Bight, Malm, Hebrides, Fair Isle, Faeroes and South East Iceland.

The general synopsis at 1900. Low Faeroes 986 moving rapidly south east, expected Sweden by 1900 Sunday. Ridge building western sea areas.

And now the sea forecast for the next 24 hours.

Viking, North and South Utsire. Cyclonic 6 or 7, and perhaps 8 at times, becoming northerly 7 or 8. Occasional rain. Moderate becoming good.

Forties	North-westerly 7 or 8. Showers. Good
Cromarty; Forth, Tyne	Northwest 6 increasing 7 or 8. Showers. Good
Dogger	Southwest 6 veering northwest. Showers. Good.
Fisher, German Bight	Southerly 5 or 6 veering northwest 7 or 8. Rain then showers. Moderate becoming good.
Humber, Thames, Dover, Wight	Southwest 5 veering west then northwest 7. Rain
Portland	then showers. Moderate becoming good.
Plymouth	Westerly 3 or 4 veering northerly 6 or 7. Showers. Good.

And now the weather reports from coastal stations for 2300.

Tiree: Northwest 5. Thunderstorm, 2 miles, 1004, now rising. (These are extracts only) Sea areas and coastal report stations do change from time to time.

The Climate

The closer you are to the equator the warmer it will be. Conversely the closer you are to the polar region the colder it will be. If you live by the sea the weather will be different to living a long way inland. A range of mountains will influence your weather. If you live in England the weather you experience will depend on where you live. In West Wales the mountains make a wet climate, the East Coast of England is drier; Scotland is colder with lots of snow in winter, the Isle of Wight is milder with little snow in winter. The Scilly Islands, 20 miles off Lands End, the most SW part of England, have a mild climate due to the warmer waters crossing the Atlantic called the Gulf Stream. If you live up in the mountains the climate is going to be cooler and more windy. How much better to live in the temperate climates of Europe and America in the summer months, and warmer climates in the winter, examples are the Caribbean. Where ever you live there must be a good rainfall. Deserts are only good for camels.

Climatic Charts

There are climate charts for every month of the year in every part of the world. Records have been made over many years and in places over 300

Forecast Plotted on a Weather Map. *((Courtesy of the RYA and R. Met Society)*

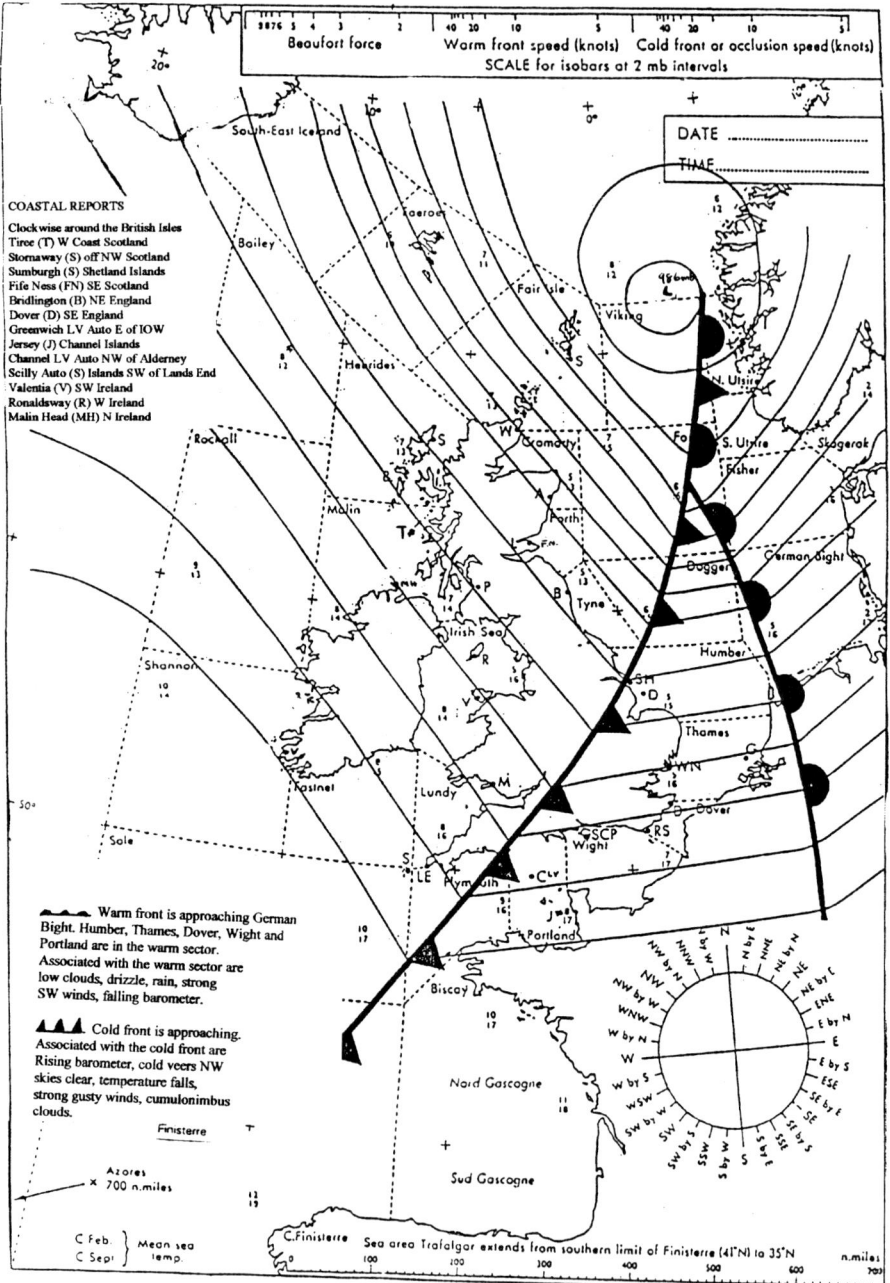

COASTAL REPORTS

Clockwise around the British Isles
Tiree (T) W Coast Scotland
Stornaway (S) off NW Scotland
Sumburgh (S) Shetland Islands
Fife Ness (FN) SE Scotland
Bridlington (B) NE England
Dover (D) SE England
Greenwich LV Auto E of IOW
Jersey (J) Channel Islands
Channel LV Auto NW of Alderney
Scilly Auto (S) Islands SW of Lands End
Valentia (V) SW Ireland
Ronaldsway (R) W Ireland
Malin Head (MH) N Ireland

Warm front is approaching German Bight. Humber, Thames, Dover, Wight and Portland are in the warm sector. Associated with the warm sector are low clouds, drizzle, rain, strong SW winds, falling barometer.

Cold front is approaching. Associated with the cold front are Rising barometer, cold veers NW skies clear, temperature falls, strong gusty winds, cumulonimbus clouds.

64

observations taken. These average values are the best record of likely weather for that month. Global warming makes these predictions less reliable, but modification to them understandable. These monthly records tell about I) average temperature, 2) rainfall, 3) sunshine, 4) barometer pressure, 5) wind direction and strength, 6) humidity, 7) surface sea temperature.

The Circulation of Weather in the North Atlantic

Warm water, and warm air in contact with it, drifts N from the Caribbean into the Atlantic. Here it meets cold water and cold air from Labrador and Greenland. Like running your bath you have to stir the hot and cold water to mix. In the Atlantic cold air undercuts warm air, setting up a depression or centre of low pressure, driven by westerly prevailing winds, these lows in winter time invade Iceland, Great Britain, France, Spain, Europe, Norway, to eventually occlude and fill, losing their identity. In winter time we see lows coming off the East Coast of North America leaving as many as one every four days. As the lows travel into the Atlantic the strong NW winds in the cold sectors behind the lows send cold air down into the Atlantic. This cold air rarely goes below latitude 20N where it meets trade wind clouds travelling in the opposite direction from Africa to the Caribbean. These trade wind clouds resemble little fluffy balls of cotton wool suspended in a blue sky, only 1,200 feet high.

CHART DIAGRAM OF ATLANTIC

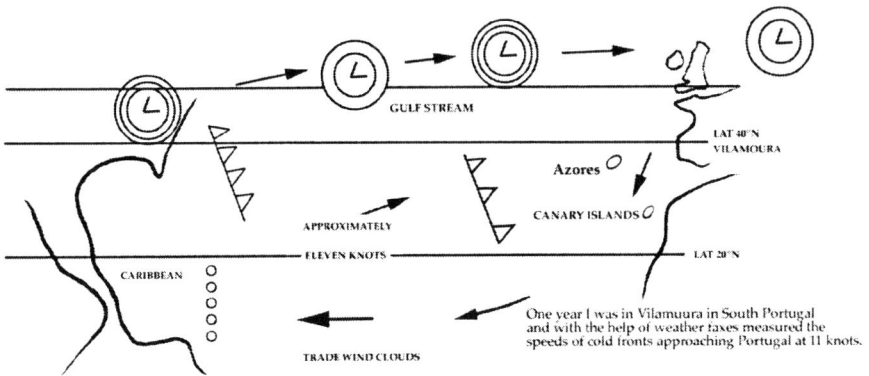

The above map is winter time. The sun is at declination 21 South, Australia is in the grips of summer time. The lows coming off the East Coast of America are causing violent swells from the north, sometimes affecting the Caribbean islands. There is a rotary clockwise circulation in the North Atlantic, the same in the North Pacific. The Gulf Stream flows in an easterly direction bringing warm water to Europe. If it fails to flow, Europe will freeze, another ice age. The trade winds blowing from the NE off the African Coast complete the circulation back to the Caribbean. Friction causes a surface drift current to flow at about half knot in a westerly direction. From December to May these trade winds vary from direction mostly NE and later in the summer to SE at an average speed of 15–25 knots. Cold air from the Atlantic tries to mix with the trade wind clouds forming troughs of low pressure, associated with rain, stronger veering winds, as they are carried to the west by the trade wind circulation. The geostrophic forces are weak and no rotation is set up, unlike the lows we have discussed.

The Change in the North Atlantic in Summer Time
When the sun starts to climb higher in the sky each day and spring flowers start to bloom, we breath a sigh of relief in England, the worst of the winter is over. The Azores semi-permanent high pressure system is beginning to dominate the North Atlantic. On page one at the bottom of the page there is a diagram of pressure systems and wind direction, showing wind blows from an area of high to low pressure. Low pressure is over the equator where turbulence and updrafts of wind produce large cumulonimbus clouds. This is called the inter-tropical convergence zone = ITCZ. Higher pressure exists above and below it, producing trade winds, from the NE and SE. On page 57 we discussed the Azores high pressure system and how this high pressure kept the lows away sometimes called 'a blocking high'. By the third week of June the North Atlantic is settled. In 1976 there was an unusual collection of high pressure areas around Europe. The continent heated and high pressure over Spitzburgen and a ridge of high pressure from the Azores over Britain met up and formed a high pressure area. This sent an easterly air flow into the Atlantic all the way across to America. A friend of mine was trying to bring a sail yacht back to England from Bermuda, and met head winds e.g. easterly winds at latitude 43°N, where there should be prevailing westerly winds.

As summer moves to autumn, rain fall in Africa has an effect on the trade wind clouds. Moisture on the African Coast causes tropical waves to form every few days. These waves move into the Atlantic near the Cape Verde Islands. These waves are areas of low pressure, carried on the NE trade winds towards the Caribbean. This is not a low as you may understand in north latitudes. The tropical wave is not spinning. The geostrophic forces at 6°N of the equator are very weak. It may start to rotate into a tropical storm, which may be triggered and intensified by storm cells from the ITCZ. The sea temperature is high and fuels the 'boiling pot'. When a tropical storm develops, it draws warm moist air off the sea into the centre of the system, causing it to rise. The water vapour it contains condenses releasing heat to the surrounding air mass causing that to rise. The system becomes self-generating. When the tropical storm reaches 64 knots of sustained wind speed it is named a hurricane. From August to the end of November hurricanes can form and reach the Caribbean and Florida. They can form in the Caribbean and outside the times given above.

WIND DIAGRAM

NORTH WIND IS VEERED
BY CORIOLIS FORCE TO
PRODUCE NE TRADE WIND

THE INTER-TROPICAL CONVERGENCE ZONE

LAT. 0⁰ ------------ EQUATOR ------------ LAT. 0⁰ ZERO GEOSTROPHIC FORCE

I.T.C.Z.

SOUTH WIND IS BACKED
BY CORIOLIS FORCE TO
PRODUCE SE TRADE WIND

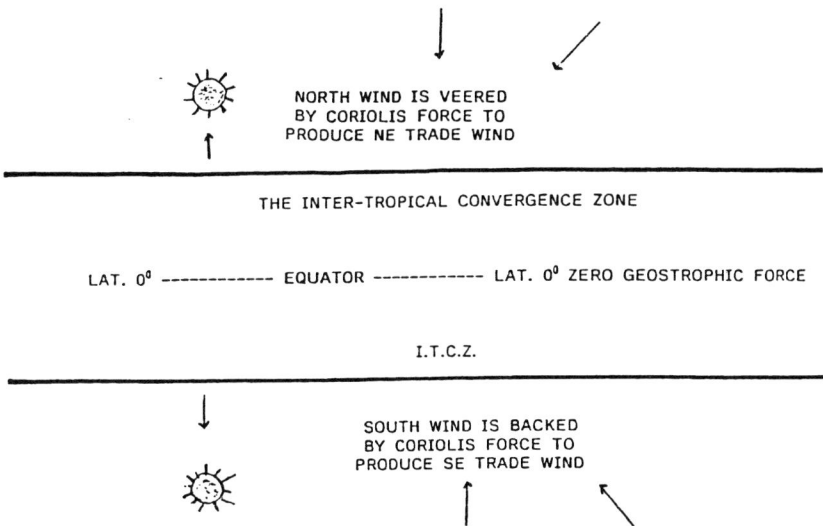

The ITCZ follows the sun, moving north in summer and south in winter. In August it reaches 12°N (sometimes higher), and moves south in October. By February it is at its lowest, 5° south of the Equator. It is characterised by unstable air and moderate to strong convection currents. When it moves north it adds to the danger of hurricane formation. If you go to St Lucia by aeroplane in September, the approach line is from NE, and the aeroplane flies around the large cumulonimbus clouds. To the south can be seen cumulonimbus as far as the eye can see, towering to 65,000 feet, way above the aeroplane at 42,000 feet.

Caribbean Weather

There are two seasons: 1) The wet season 1st June–30 November when hurricanes can form. 2) The dry season 1st December–31 May. With global warming and more rain the wet and dry seasons are becoming very similar. Trade wind can become lighter and variable in summer months.

Conditions for a tropical storm to develop are:

1 A tropical wave – A non-circulatory area of low pressure.
2 Sea temperature above 26°C.
3 Convective activity. A high island like St Vincent 4,000 feet adds convection. Also storm cells from the ITCZ.
4 Favourable upper level wind and atmospheric conditions.
5 Rotational forces.

Hurricane movements: Early hurricanes tend to form east of the islands, and travel at a lower latitude. Late hurricanes can form west of St Lucia. Trinidad hasn't had a hurricane in 50 years, so many yachts lay up in Trinidad. 85% of hurricanes go north of St Lucia at latitude 12 N. Dominica, Antigua, St Barts, Virgin Islands are most often hit. Late hurricanes from Cape Verde Islands move NW and lose their intensity in colder water. The Azores and Canary Islands have been hit by very bad storms in 1998. Hurricane movement is very difficult to predict, especially slow moving systems. The sea surge can be large as a hurricane goes past. What was a sheltered harbour becomes exposed.

THE SAFFIR – SIMPSON SCALE

Category	Pressure	Sustained winds	Surge
I	980 Mb	64–82 knots	4–5 feet
II	980–965 Mb	83–95 knots	6–8 feet
III	965–945 Mb	96–112 knots	9–12 feet
IV	945–920 Mb	113 134 knots	13–18 feet
V	< 920 Mb	> 134 knots	> 18 feet

Hurricane Floyd in the Caribbean 14 September 1999.

Hurricanes are recorded on their strength, barometric pressure and sustained winds, producing a likely sea Surge, Called the Saffir–Simpson Scale.

Jet Streams

These are fast moving air currents up to 12 miles high, often much lower, which circle the earth in both north and south hemispheres. Their position is usually near polar fronts but they can change rapidly and dip down over lower latitudes. They can bring down cold air, and small depressions tend to deepen and leave at their 'exits'. One of the effects of global warming appears to make them behave more erratically, dipping down towards Florida and the British Isles. When they come down the winds are stronger. When they go up to 100,000 the winds become light. They can be 1,000 miles long, 3 miles deep, and up to 300 miles in width. Speeds can reach 200 knots. When they dip down to 30,000 feet, they blow the tops off the cumulonimbus clouds.

El Nino

Hot vents and underwater volcanoes in the Pacific Ocean cause hot water and hot air in contact with it to drift to the West Coast of America. The fishermen complain the fish have gone away. The hotter air drifts into the Caribbean, increasing Caribbean air and sea temperatures. The air pressure rises more in some areas and falls in others. Air travels on the pressure gradient, from high pressure to low pressure. Thunderstorms travel on the gradient wind. El Nino caused the high pressure over Florida to increase producing a pressure gradient, from 1036 over Florida to 1012 over Trinidad. On 10 October 1998 it rained heavily for eleven days in St Lucia. A trough of low pressure set up a circulation and as it moved westward 500 miles west of St Lucia it formed Hurricane Mitch on 22 October. The hurricane moved very slowly, only 5 knots and wind speeds were
    recorded at 150 knots for three days. It was expected to move NW to the Yucatan Channel. It didn't. It moved due south, sank the 3-masted sailing

ship Fantome with the loss of 31 crew, and moved into Honduras and Guatemala, killing 10,000 people, before clearing the Yucatan peninsular, clearing Florida and arriving off the North Coast of Scotland. It had travelled on the gradient wind, the pressure in Florida 1036 Mb, the pressure in Trinidad 1012 Mb.

1997 was a strong El Nino year, which dried up the African Coast and there were no hurricanes. A jet stream came down to 33,000 feet and blew the tops of the cumulonimbus clouds, so further convection didn't take place.

El Nino has the effect of causing the jet stream to wander off course, dipping down unexpectedly low, and causing cold air to come to lower latitudes down from Canada.. Where cold air meets warm air masses, terrible storms occur.

La Nina is the name given to these colder effects. Terrible storms on 2 January 1998 caused outbreaks of extreme weather, snowfalls, mid-winter tornadoes and cold in America.

JET STREAM DIAGRAMS

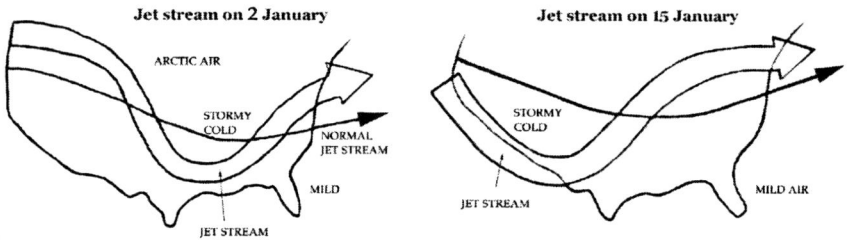

Evidence of Global Warming

The impact of global warming is felt worldwide. The dipping down south of the jet stream over Britain is likely to produce worse winters, worse storms, and drier, hotter summers. Evidence of global warming can be seen elsewhere. Increase in sea temperatures in the Pacific and Indian Ocean have caused coral reefs to die. The coral reefs are 'bleached'. Half of the coral reefs in the Indian Ocean are damaged. At Point Barra in Alaska ice is breaking up 2 weeks earlier than 20 years ago. Polar bears have less time to catch seals and mothers have less fat to give to their cubs. Ice is melting in the South Pole. Penguin numbers have decreased. Migration of birds is taking place earlier. In America sand hill cranes which evolved in dinosaur times are migrating 20 days earlier than 30 years ago. In rain forests in South America there are 50 species of frogs and toads. Many species have disappeared in the last 30 years, including the 'Golden toad' unique to Monte Verde. Meteorologists say the number of dry days in Monte Verde has increased. The cloud base on mountains has moved upwards. In Europe 22 types of butterflies have moved north. Sea level is rising.

The Earth's Temperature Controls are:-
1) the sun, 2) the oceans, 3) the atmosphere, 4) volcanoes. The sun produces fluctuations in the form of solar fares, but these are not causing an increase in earth's temperature. The oceans cover two-thirds of the earth's surface. An increase in temperature causes more evaporation and more precipitation. Also it affects fish distribution and everything that grows in it. Any alteration in ocean temperature causes an alteration in the air in contact with it. The atmosphere must be free of pollutants to avoid the 'greenhouse effect', or in the case when sunlight is totally blocked from reaching the earth, cooling and ice age.

Volcanoes – there are 1,000 active volcanoes. If large eruptions occur there will be atmosphere pollution. If several large ones erupt in the same period, there will be global cooling e.g. in 1902 Mount Pelee and several other eruptions caused a fall in earth's temperature of one degree. The same happened when Krakatoa erupted in 1873. In 535 AD Krakatoa erupted with enormous force. It is estimated 15 miles of Krakatoa Island disappeared in I second into dust, so fine it wouldn't settle back to earth. It took three months for the dust to engulf all latitudes and then it blotted out the sun for five years. The earth's temperature fell, crops died because photosynthesis was much reduced. Half of the world's population died from starvation and the plague. The fine dust did eventually fall back to earth, but caused terrible trouble. It blocked the stomachs of the fleas that live on rats and transmit the plague. The fleas developed an incurable appetite and multiplied in vast numbers, attacking cattle and humans. It took 40 years for the earth to recover from the atmospheric pollution. Forest fires cause atmospheric pollution e.g. the vast forest fires in Indonesia in 1999; also industrial burning, car exhausts, aeroplanes, etc. So volcanoes are the principle cause of temperature control by heating the oceans, heating the air in contact with the ocean, and polluting the atmosphere with debris from eruptions, and the greenhouse effect. Periods of global heating and cooling occur. Meteorologists have not kept records long enough to be able to tell if these present fluctuations are within normal limits or whether there is a

THE PATHS OF SOME HURRICANES

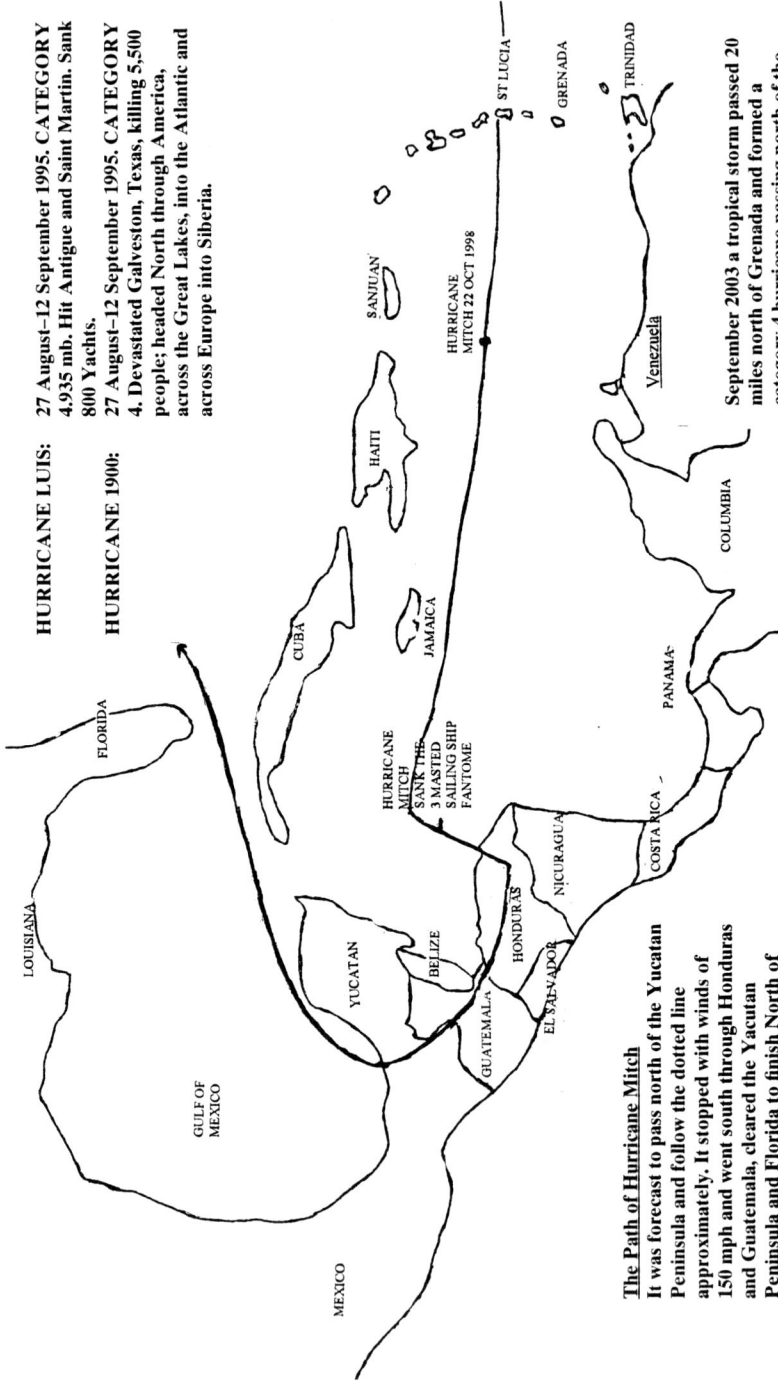

HURRICANE LUIS: 27 August–12 September 1995. CATEGORY 4.935 mb. Hit Antigue and Saint Martin. Sank 800 Yachts.

HURRICANE 1900: 27 August–12 September 1995. CATEGORY 4. Devastated Galveston, Texas, killing 5,500 people; headed North through America, across the Great Lakes, into the Atlantic and across Europe into Siberia.

FLORIDA

LOUISIANA

GULF OF MEXICO

MEXICO

YUCATAN

GUATEMALA

BELIZE

EL SALVADOR

HONDURAS

NICURAGUA

COSTA RICA

PANAMA

CUBA

HAITI

JAMAICA

SANJUAN

ST LUCIA

GRENADA

TRINIDAD

Venezuela

COLUMBIA

HURRICANE MITCH SANK THE 3 MASTED SAILING SHIP FANTOME

HURRICANE MITCH 22 OCT 1998

September 2003 a tropical storm passed 20 miles north of Grenada and formed a category 4 hurricane passing north of the Yucatan Peninsula into the Gulf of Mexico and going ashore in Louisiana. A few weeks later another hurrican followed the same path.

The Path of Hurricane Mitch

It was forecast to pass north of the Yucatan Peninsula and follow the dotted line approximately. It stopped with winds of 150 mph and went south through Honduras and Guatemala, cleared the Yacutan Peninsula and Florida to finish North of Scotland. It had travelled on the pressure gradient.

need to take action, and what action to take. I would rather live in a period of global warming, provided I don't get boiled, than live in the next ice age chasing polar bears around the South of France. The sea level is rising at the rate of 1½ feet per 100 years. In the Scilly Islands, off Land's End, England, the sea level has risen 30 feet in two thousand years. Will the rate of rising water levels increase? Sea defences are expensive e.g. keeping water out of London.

When ice bergs melt they only displace the same volume of water. When ice melts over land this will cause sea levels to rise. The sea temperature is increasing in the North Atlantic and is one degree higher. If the Arctic ice melts how will this fresh water effect the sea? Will it cause the Gulf Stream to move south toward Spain? If so, we can keep our polar bears on the permanently frozen Thames. If we could predict the future could we take the right action? In November 2000 there was a yacht race from the Canary Islands to St Lucia in the Caribbean (an annual event). After a bad start with strong head winds, the trade winds blew from behind, and all appeared

World Currents

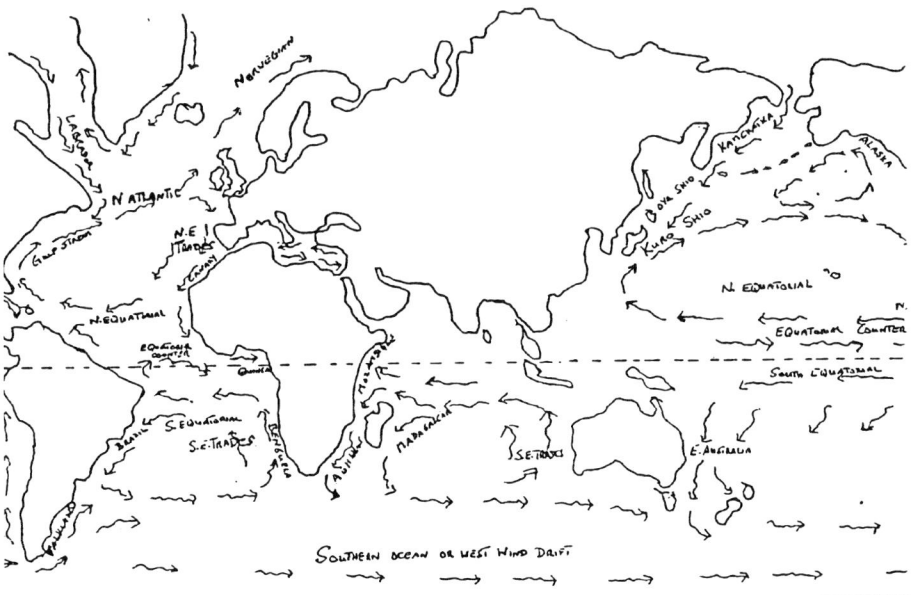

to be going ok. Then the wind from the east stopped and a westerly head wind arrived. Eventually the westerly head wind stopped, and left 250 yachts in mid-Atlantic with no wind, until eventually the trade winds came back. What had happened? El Nino was to blame. A late hurricane had formed in the Caribbean and with global warming this altered the pressure distribution. So Hurricane 'Lenny' travelled on the pressure gradient 800 miles due east, in the wrong direction, before filling and stopping. It sent a large swell crashing into the normally sheltered side of the Caribbean Islands, causing property damage. This swell was predicted and reached as far as Trinidad.

Global Warming

It is happening. It will involve rise in sea levels, more rain, more flash floods, mud slides, worse storms as more extremes of hot and cold air try to mix; more tornadoes, damage to Arctic and Antarctic life as more ice melts; damage to species unable to cope with changing habitats; alteration to the jet stream, producing areas of intense cold. We will have to learn to adapt. Perhaps the slogan will read – 'buy a mobile home and migrate'. What will happen if the north and south poles melt? Will we have a global alteration in world currents?

Chapter 4

The discoveries of early man
and his beliefs

Why do we start looking *for* clues 150,000 years ago when early man was present one million years ago, and we were almost identical to chimps five million years ago? The climate and genetics provide the answers, The human brain was well formed 150,000 years ago. Why did this happen? We are told the reason lies in communication which leads to reasoning and the ability to plan ahead. What communications were there with no written language, no education and no schooling, and nomadic people living in small groups with a low expectancy of life. Are we to believe that the chemistry of neurone connections had reached such a level, the powers of reasoning begin to form? Why didn't a chimp develop the powers of reasoning? A baby chimp displays all the features of a baby human. If we are to believe the reason for a big brain so long ago, we must have inherited it from a superior race 500,000 years ago, which became extinct due to some earthly disaster. I discuss these possibilities in my chapter on the human brain. At this stage accept the fact we have inherited a large brain which up to now hasn't been developed, and used to its full capacity. Now we can start the story from 150,000 years ago.

There was an enormous ice age 150,000 years ago. Man would not have survived in this prolonged period unless he had migrated south into Africa. The seas were frozen over in places and the sea level fell by about 300 feet.

Genetic data only provides an approximate date, but combined with climatic changes and archaeological findings gives us more clues. Fossil remains have been found in Africa from 125,000 years ago. Primitive tools were found, stones made into sharp edges for cutting. It is estimated the human race had nearly become extinct and there were probably only 10,000 people alive. The science of genetic origins show we can trace our origins through women. Mitochondrial genes are passed on by women. This means we have inherited Mitochondrial genes from an original women or African Eve, because these genetic lines are the only ones that have survived today. This doesn't answer what had happened say 300,000 years ago.

We don't know the answers to this period. Why did man need to migrate north from Africa 80,000 years ago? The answer comes from the changes in climate. The ice age caused other effects besides cold and a fall in sea level. The land in places became desert. Man needed to migrate to find food. Early attempts to migrate travelling north would have been difficult with desert regions to cross. The latest findings suggest this migration occurred across the southern end of the Red Sea at its narrowest point, with sea levels 150 feet lower, the crossing would only be a short distance from Ethiopia to the Yemen. The migration continued into Iran, probably by the Straights of Hormuz which may have been joined to Southern Iran. Beach combing would have provided fish for their diet. Migration continued into India.

Reduction in genetic line occurs when groups live in small numbers, resulting in only one genetic line surviving. We all have a common ancestor. When large numbers of people live together, more genetic lines occur. Genetic findings indicate the first Homo sapiens went to India.

The migration continued to Malaysia and the tropical rain forests. Homo sapiens learned to live in the tropics, with better food and fresh fruits. The biggest volcano for two million years erupted in Malaysia 74,000 years ago. The fine dust and ash would have blotted out the atmosphere for years. Excavations reveal stone tools used by these people buried in the volcanic ash. Some inhabitants survived the volcano and they have their own genetic lines. Later these Homo sapiens migrated via the Indonesian Islands to Australia.

How can we discover our ancient history if there are no written records? Can we imagine what it was like 50,000 years ago? Some systematic approach to the problem must be followed along the lines of scientific analysis and every avenue of research. Let's start with man's limitations, especially his understanding of his environment. There was no written language, no teaching, no schools of learning. Life was nomadic.

I Will Try to Explore the Evidence Under the following Headings
1 How the brain 40,000 years ago may have reacted to his environment.
2 Evidence of a hieroglyphic script.
3 A study of people from different parts of the world, studying their pyramids, temples, analysing their astronomical and religious significance, studying ancient philosophers, ancient gods, legends, tracing biblical stories and archaeological excavations.

4   The land and sea as a means of communication, trading and the spreading of knowledge worldwide.
5   The influence of religion.

1 The brain was fully formed 40,000 years ago. It was well developed 150,000 years ago. Why it developed into such a large complex structure is a mystery. The love of art and music and creative achievement, coupled with deep emotions seem second in importance to survival. It can be partially explained away in the bonds of sexual activities leading to a dependency and desire to foster our children. The remainder is explained in the development of language, communication, the written word, memory patterns and the ability to plan ahead, which animals have very limited ability. The brain needs several generations of learning to be able to analyse the incoming signals and respond accordingly. The process of knowledge is a step by step accumulation. 100,000 years ago man was roaming Europe, Asia and probably America. Life was nomadic, travelling in small groups, making tools for hunting and fishing and coping with cold. Records discovered in caves 30,000 years ago indicate there was a start in art forms and pottery. There was no written language, communications primitive, and 'development' as we understand it today, very limited. The brain was fully developed but not being used to its full advantage. Let's dispel the myth that the human brain developed into its fully formed state 100,000 years ago as a result of a written language, education and advancements in scientific discoveries.

Homo sapiens may be questioning the existence of himself in the environment and the place played every day by the sun, moon, planets and stars. Would he have understood the colours of the rainbow or attributed it to a rainbow god? What would he have thought of a shower of meteorites lighting up the night sky like fireworks. What message did it convey? Perhaps it was a god, or several gods. There may have been a superior race 500,000 years ago accounting for the development of the brain, but I cannot find any evidence. I believe the brain developed to its present size for reasons unknown, and now it is up to us to use it, increase the memory storage and the density of neurone connections.

2 Hieroglyphic Scripts
There is no evidence to suggest Homo sapiens had any written words 100,000 years ago. Probably 40,000 years ago there were the first signs of communication by pictorial drawings of birds, animals and people, and

these pictures having a meaning. Other symbols would be formed to transfer a message, and knots tied in cord indicated the early counting of numbers. This became necessary at the start of trading, and the weighing of gold (a universal early currency). There were 24 of these marks which were referred to as the ancient Egyptian alphabet. The Mayans in America had also developed an alphabet. The Indus civilisation in ancient India had a hieroglyphic script dating from 2500 BC. They used hieroglyphics on seals on merchandise.

When did the communication from speech to a written language take place? The answer is about 6,000 years ago, about 4000 BC. This is about the time the Sumerians developed parchment and paper. I think the Chinese had invented parchment prior to this date. The earliest Egyptian writing was called hieroglyphic script which is found on the walls of pharaoh's tombs, ancient temples and mummy cases. Early glyphs were animals. Producing two glyphs together would produce a sound but it would bear no relation to the original meaning. These sound glyphs are called phonograms. Hieroglyphics are fundamentally independent of language and would have allowed inter tribal communication over long period of time while language differentiation developed.

3 Lets start in Europe.

Homo Sapiens Neanderthals occupied Europe and Western Asia as long ago as 100,000 BC. They probably came from Africa. By 40,000 BC they were reported in Europe, living in small groups of 8 to 25 and were very adaptive. They lived in caves with animal skins for clothing, hunted with spears, made fires to cook and keep warm. Their diet was 85% meat, their bottom jaws were more developed to cope with their diet than today with our soft foods. It was now the start of the ice age and by 25,000 BC winters were bitterly cold in the South of France and dangerous animals like bears were roaming. Only 20% reached the age of 40. They had developed a language of their own and developed art forms. They spread across Asia eventually reaching Australia.. The harsh winters of the ice age took their toll and their numbers dwindled. They were replaced by other people called Cromagnons who came from Africa. They had a more sophisticated language and organised themselves in larger groups, outnumbering the Neanderthals and taking over their caves and dwellings. There are no genes in modern man suggesting Neanderthals survived. The Cromagnons made clay pottery and small statuettes. Cave paintings in France and Spain are dated 25,000 BC.

With the retreat of the ice age in 10,000 BC man had a better chance of staying in one place, building a home and learning to cultivate crops. Towns grew up on the Euphrates river and fertile Nile valleys, metal workings began, pottery, artwork, spinning wool, bread making, storage of wheat and food, and trading.

Approximate dates are:

7,000 BC	First towns built on the banks of the Tigris and Euphrates. And banks of the Nile (near Cairo).
5,000 BC	Towns built in Mexico, China, Russia and India.
5,000 BC	Central America, Maya people built cities with large pyramids and temples. Teotihuacan had over 100,000 people living in the city with hundreds of pyramids.
4,000 BC	Pyramid temples in Peru.
3,500 BC	City of Urux built on the Euphrates, a river in Southern Mesopotamia.
2,800 BC	Stonehenge.

2,686 BC–2,181 BC 3 pyramids of Giza in Egypt.
Sphinx probably built 13,000 BC

It is suggested the sphinx is much older than the pyramids, and one method used in dating is water erosion. The climate was different and wetter.

Gold and ivory became great ornaments and art treasures were required by the rich for accompanying the deceased into the afterlife e.g. Tomb of Tutan Khamun. This inequality in wealth led to the very rich getting more sophistication and refinements and has provided the way ahead in the development of ideas, a good area for research.

The pyramid and temples were both burial chambers and astronomical sights for observing the sun, moon, planets and stars. At the same time of the Great Pyramids, 25 other pyramids were built along the West Banks of the River Nile, near Cairo. The three pyramids of Giza are aligned on the ground as stars of Orion are aligned in the sky. We know from history the ancient cities had rulers, Kings (pharaohs) and priests, and other gods and idols to worship. Much time was spent in preparing royal burial chambers to ensure the kings have prepared themselves for the afterlife in immortality amid the stars.

The Great Pyramid is massive. Its sides measure 700 ft. Some of the stone blocks weigh 40 tons, a great feat of engineering and gigantic organisation to feed and instruct so many thousand workers. The pyramids line up with the geographical axis of the earth and star shafts attract the light from certain stars. The construction takes into consideration precession of the stars. The relationship between the stars and the pyramids is that they only match the ground pattern in 10,500 BC due to precession. What is precession? The moon and the sun pull on the earth's equatorial bulge causing the rotational axis of the earth to trace out a cone so that the North Pole which points towards the North Star of Polaris at present will move away from the point and trace out a cone of precession returning to point at Polaris in 26,000 years.

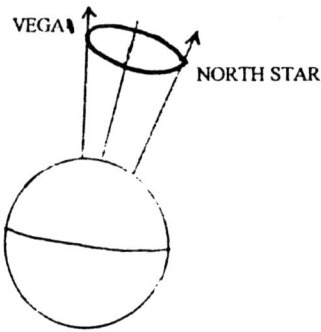

This mathematical knowledge was known before these pyramids were built, and also known independently in America. The angle changes one degree every 72 years. The first known star charts were produced at a much later date in China and Greece about 200 BC and probably in Mexico.

Ancient gods and philosophers in Egypt – An understanding of their beliefs at this stage may help us. Atum was the very first GOD, the product of Divine Magic. Existence from non-existence came at a preordained moment, causing something to emerge out of nothing. According to the ancient book of the Pyramid Texts, a primeval mound expanded into a high hill and Atum materialised, and from heaven fell a triangular stone, the Benben stone (sounds like meteorite) carrying secrets of the Duat. The Benny bird (= heron) appeared perched on top of the primeval mound. The Benben stone was once displayed in the temple Het Benben at Heliopulis (the city of the sun). Ancient texts tell of the history of the beginning. God Osiris (identified with the star Orion) brought civilisation to Egypt and with the great Goddess Isis (identified with the star Sirius) produced a son Rorus (portrayed in the temple Dendera). The God Isiris became lord of the Duat to act in judgement of the souls of the dead. Departed souls could attain immortality in the Duat, an area in the sky called Orion. The precise alignment of the burial chambers and the pyramids could ensure the sky was correctly aligned on the ground. The importance of being able to go down from any sky they choose is stressed many times over. It was believed everything was made in the sky and was brought down to earth. Troth was the ancient Egyptian God of Wisdom. Maat was the Egyptian Goddess of truth, justice and cosmic equilibrium. There were 42 books of Troth containing all the knowledge of the world. Khufu ruled from 2551–2528 BC, the builder of the Great Pyramid of Giza. He searched for the secret archives of the sanctuary of Troth for the books. A son of Rameses II who ruled 1290–1224, said he had found a book of Troth and a great light shone from it. It contained the secret knowledge that allows the soul to pass the Duat to immortality between Orion and Sirius. In 1500 BC there was a great astronomer who claimed to know the wisdom of Troth called Senmut, and to have all knowledge since the beginning of time. He must have influenced the thoughts of Tutankhamun who ruled at a later period 1333–1323 BC. In the second shrine of Tutankhamun there were tablets showing the importance of the connections made in thought between the brain and the stars (also called – keeping the bridge open to the stars).

This bridge to the stars we find in the beliefs in the ancient Americas and much later in the Incas. The ancient city of Teutihuacan in Mexico built great pyramids, with the same astronomical and philosophical thinking i.e. making ground patterns of the stars to ensure the souls of the deceased entered eternal life.

Ancient beliefs told us of a God of Fire that heated us from the sun, The Sun God Ra, a Moon God Nanna, and many other beliefs. The Ziggurat at Ur, built by King Ur Nammis (2112–2095 BC) symbolises the Moon God Narma. The ancient philosopher Galen (190 AD) was a remarkably knowledgeable man. He said the circulation of the blood was like the ebb and flow of the tide. What a pity Galen did not have a microscope to see the blood went all the way round and back to the heart. This suggests research into how things worked had been going on for a long time. The golden age of knowledge must have been 300 years from 600 BC to 300 BC. This produced a great many scientists and philosophers.

Nebuchadnezzar   604–562 BC (made the hanging garden of Babylon)
Herodotus   484–425 BC
Socrates   469–399 BC
Plato   427–347 BC
Aristotle   384–322 BC
Alexander the Great 356–323 BC

Aristotle was a pupil of Plato and taught the son of Alexander the Great. He spent his entire life in study and teaching. He taught the earth was the centre of the universe. Everything went around us. This concept gave us a sense of great importance. The gods had created life for us. How could we contact them? Aristotle had formulated ideas on about everything, including a system of logic. Philosophical thinking suggested the brain reacts to the world as a result of its form, which is unreal and an illusion. The working of reality was understood by much learning involving the higher part of the brain, the soul. This view was held by ancient men in India, Cambodia and Mexico.

So I have decided to have a dinner party and I am going to invite Socrates, Herodotus, Plato, Alexander the Great and Aristotle. I must prepare the roast goat and trimmings! 'Gentlemen, I must introduce you to our findings in the year 2003. The earth is not the centre of the universe. The earth goes round the sun, and our sun is only one of billions of stars. All elements are made in the stars, and we are made from DNA. Also we cannot agree with an all important Omnipotent God and loving god concept and have at the same time all the pain and suffering on earth. Can I invite a reply?' Aristotle stands up 'God obviously made you and your DNA and then left never to return. I thought of this first 2,250 years ago'.

May I Ask You Another Question?
Modem techniques allow us to study the human brain. In the last 100,000 years the human brain has become large. There seems to be no reason why we should have developed a large and complicated brain which was fully formed as we know it today 40,000 years ago. We believe that this development came about by survival, hunting skills, adaptation to the environment and genetics. The skills we have developed with our hands, the written language, understanding to grow crops, to build houses for shelter, to develop large communities, the need for sociability, all moulded our brains. Why do we have a music centre in our brain? Is this part of our survival? Why did we develop a large brain for reasoning and planning? Animals have not had the need to do this? Aristotle replied:- The brain has two functions:- the function of operating the body organs, eating to fuel the furnace within us by which heat and nourishment reach our body organs, and waste products are disposed of, and the higher centre of the brain the PSCHE which consists of two parts:- part one is the development of characteristics acquired by habits (Greek Ethos); Part two is the strive toward complete happiness, which varies from man to man, but emphasises

that the ERGON (activity) demands of a man that he strive to make great achievements. This forms his total PSCHE which passes with him into the next life. This is enhanced by wisdom to understand intellectual excellence and a good moral code. In achieving these higher states the partition between the next life and this comes closer. It is that the god who made us, may have made us in a similar image and wants us to develop to higher levels. Communication to GOD through the PSCHE has altered our brains to this larger size, so GOD likes MUSIC'.

Gentlemen, this is a most interesting dinner party. Can you tell me what evidence you have found for ancient civilisations?

Plato replied to my question. We have a history from the beginnings of time given to us by the Gods. There are texts, some of which we cannot translate, dating from very ancient times. An example of the Pyramid Texts dated 2350 BC is carved into the walls of the Pyramid of Unas, near Giza. The coffin texts are available for inspection in Egypt and date from 1800 BC; also the book of the Dead of Userhat from about the same date. The 42 books of Troth have been lost.

The great ice age caused people to migrate south in search of warmth. When the ice started to melt, sea levels started to rise and much land was covered by water. Everyone had to retreat to the mountains. All fertile lands were engulfed. When the water levels receded, everyone decided to live in the fertile places and large communities came together.

I told Plato we would confirm the dates as the last ice age started to melt about 12,000 BC and was complete 8,500 BC when the sea level rose 400 feet.

Ancient texts in Egypt worth mentioning are the book of the dead, the pyramid texts, the book of the gates, and the book of what is in the Duat and the Books of Troth. How exciting if we could uncover some of these writings, the 42 books of Troth and even find the Benben stone. The pyramid texts were so old in 2300 BC that they were not fully understood by the scriber of the day, and they had been copied from much older texts. There are no hieroglyphics on the walls of the great pyramid at Giza, but there are some on the many other temples near Cairo and on the banks of the Nile. There are unexplored tunnels under the sphinx and one un-explored chamber in the Great Pyramid. One day the Egyptian authorities may allow further research to these areas to take place.

In the Aegean a truly remarkable civilisation grew up in Crete. It was also a source of ancient religious traditions. The period was 2200–1450 BC. It was mentioned by Homer in his Odyssey, by Aristotle, by Herodotus and many others. King Minos had acquired from trading great wealth and a powerful navy. Palaces were built at Knossos in the north, Phaistos in the south. There developed a great social order and style of living with modern ideas of flushing toilets and showers. The palaces were elaborately decorated. They did not speak Greek but developed their own language which could be written as early as 2,000 BC. Later the written word became based on an alphabet. The Minoans inhabited other islands in the Aegean including Santorini, an island nearby. They played games including a daring one called 'Somersaulting over the charging bull.' In 1924 the archaeologist Sir Arthur Evans uncovered the palace of Knossos and excavated the ruin. Poseidon was the God of the Sea and Earthquakes. In about 1450 BC the volcano on Santorini Island, 75 miles away, erupted with enormous force. It is estimated the explosion and what followed was four times the size of Krackatoa in 1883. The pyroclastic flow, volcano ash, and tsunami that followed destroyed Crete. The Cretan navy was lost and the Minoan empire disappeared, probably because they had lost their headquarters, their navy and they did not speak Egyptian.

Trade routes opened up to the far east of Asia, to bring back silk from China which was much sought after by the rich. China has had 8,000 years of civilisation. Their country was in many states each with their own language and cultures. The Shang dynasty was 3,500 years ago. Eleven tombs devoted to the eleven rulers have been found. In the third century BC Qin Shi Huangdi conquered these states and organised them into one state with one language. He was responsible for the tomb of 8,000 life size terra cotta soldiers, which were made on a production line. The secrets of paper making were brought back.

There were two great temples in Cambodia, Angkor Thom (= Angkor The Great) and Angkor Wat (=Angor The Temple). The temples viewed from the air give the impression of a massive fortress, a moat, an outer wall, an inner wall, a central high area. What secrets could have been kept locked up; perhaps books of Wisdom, Gold, or a Great Diamond, maybe the Crown Jewels? The precise astronomical alignment of the temples shows a knowledge of precision. The exact alignment of the stars projected on earth at Angkor matches the position of the stars of Draco at the sunrise in the spring equinox 10,500 BC, i.e.. the knowledge of how to go down to any sky appeared to be universally known. Again at Angkor we see an enormous feat of engineering, knowledge of precision and an astronomical observatory, a religious site where the souls of the dead are directed into immortality, and a fortress to defend something or somebody. The knowledge of precession had spread from Egypt. Could it have spread across the Atlantic to America, the knowledge to design and build pyramids, precisely aligning them to the position of the stars as they would have been on the dawn of the spring equinox 10,500 BC. The number 72 comes into a lot of alignment projects and distances.

In the East China sea to the East of Taiwan lies an island Yunaguni. Recently a diver found what appears to be a man made structure, a part of a monument, at depths down to 100 feet. This is accurately aligned North South/East West. Some of the blocks are 200 tons. The sea level has been rising at an average rate of 15 feet per 1,000 years since the melting of the ice caps 10,000 BC. For this monument to have stood 50 feet above sea level would suggest the age at 10,000 years old. It could be argued the land had subsided and that the huge blocks were shaped from similar natural shapes. Astronomical considerations suggest a similar age to 10,000 years ago.

Easter Island in the remote part of the Pacific Ocean has many statues, some of large size. I believe these were erected to remain in contact with deceased ones. The inhabitants came from neighbouring Polynesian islands, and to me it belongs to a curiosity factor of more recent times, not that of an ancient civilisation.

In the Hindu religion the great creation was by Vishmu. In South America, Andean history tells us the earth was created from lake Titicaca by the God Viracocha. In Mexico city the ancient teacher was Quetzalcoatl. Mexico city stands on the remains of a previous ancient settlement Tenochtitlan. Some 100 miles to the North East of Mexico city there was

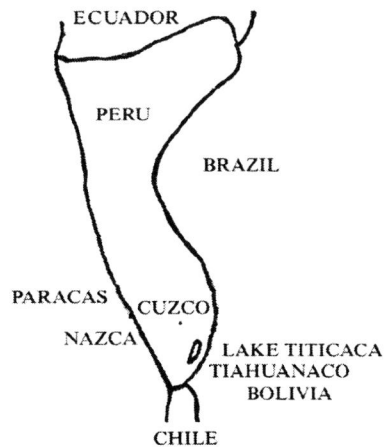

another ancient city Teutihuacan, with many pyramids. Quetzalcoatl claimed to have died and come back from the dead. Legend said that he lived at a time when everything in the world could be viewed without moving place. It sounds like the out of body experience of the Hindu when they can leave their body to go on a long journey. Today it sounds like satellite communication system and worldwide TV.

On the west side of South America is Peru. Midway down its coastline is Paracas and 125 miles South Nazca.. Geoglyphics is the study of ground drawings. Some of the most famous are the Nasca lines. Other Geoglyphics exist close by, The most famous one the 'Candelabra of the Andes'.

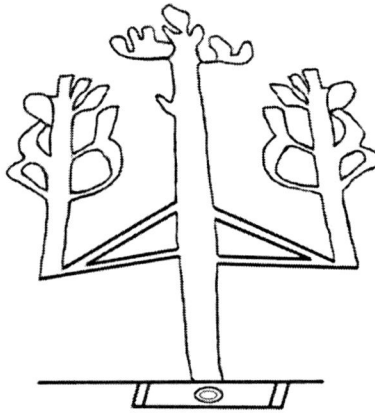

Candelabra

Whether this is part of the Nasca lines nobody knows. It looks like a space craft landing. The candelabra could represent a path of enlightenment, a scene from the sky projected onto earth. The Nasca lines have been studied by Maria Reichi for many years. They appear to map out the constellations on the earth as well as the drawings of birds, monkeys and spiders of enormous size to be viewed from space. The idea of bringing the sky down to earth was a very ancient view and would enable the people to walk amongst the stars like a pilgrimage to maintain the bridge to the Gods. The candelabra and Nasca lines are about 2,000 years old. It is not impossible Andean people could have developed a large kite so they could have viewed their artwork in the Nasca lines from the air. Such kites were known in Tibet. Some of the Nasca lines look like aeroplane runways. But we must be careful of interpreting old ideas within the wrong frame of knowledge.

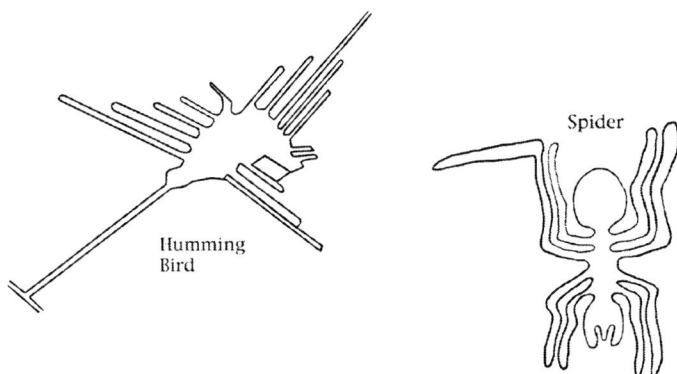

Nasca lines

Another *350* miles to the South East lies lake Titicaca and the ancient city of Tiahuanaco, from whence time began. There are megalithic structures near Cuzco, the fortress of Sacsayhuaman hill and at Tiahuanaco.

Ancient legends tell the story of the great Quetzacoatl and his teaching, with pale skin, who came down to earth to teach the people, for he knew everything since the beginning of time. The teachings centred on peaceful existence and knowledge. When Quetzacoatl left it was said he would return at a later date. But things were not to be, for a barbaric race of men, moving South from North America, infiltrated Mexico City, and took control. The peaceful existence was gone. They ruled by fear, created by themselves for political gain, and abuse of wealth and privilege. While displaying feelings towards the doctrine of Quetzacoatl, they branched out into human sacrifice in a psycho phrenic display of evil dictatorship, getting their wires badly crossed. In 1100 AD these Aztecs were in control. When the Spaniards came in 1590 they were horrified by the human sacrifices they saw. The Aztecs thought Quetzacoatl had returned, but no, the Spaniards came to loot gold, for personal gain, and murdered the Aztecs and later the Incas. Sadly they had no regard for history and burned the library of books telling us of ancient civilization before the time of Aztecs and Incas. They did this in the name of the Roman Catholic Church. Religious fanatics and fundamentalists of all religions are still attempting to destroy anything which contracts their ideas.

The Mayan people had arrived in America a long time before the Aztecs. They said there had been 4 previous epochs, when catastrophes had wiped the face of the earth and the 5th epoch started on 13 August 3114 BC and would end on 23 December 2012, due to a great movement of the earth. The priests and followers of Quetzacoatl are recorded in the sacred book of the ancient peoples Maya of Mexico, called Popul Vuh. The texts refer to a golden age when men knew everything. The pyramid at Chichen Itza on the Yucatan Peninsula is like an astronomical observatory. It is of great antiquity.

Questions are asked about date's, e.g. did the ancient town of Tiahuanaco, exist before the Greek and Egyptian counterparts? We do not know. The cave paintings in the Lascaux region of France have been dated at 25,000 BC, and some interpretation suggests an attempt at drawing stars. Maybe some temples are much older than we thought. The first human, Neolithic man, probably reached Alaska from Siberia 80,000 years ago. Records have been found of people living in the Arctic 20,000 BC. In 12,000 BC there are records of people in Chile and moving South to the Andes. People moved into South America, into the Andes about 10,000 BC. The Incas came much later, and their large empire stretched from Peru, Chile, West Argentina, Bolivia, Ecuador and part of Colombia. The Inca town of Cuzco was highly developed by 1400 AD. The Incas expanded their beliefs they had inherited. By going high into the mountain they could observe the stars and shapes made by gas clouds. The conclusions they reached led to the horrors that followed – the importance of keeping a bridge to the stars resulted in an offering to the Gods, the sacrifice of children high up in the mountains to please the Gods. They even tried to foretell the future, and their own destiny. They believed they were a superior race and did control a large coastal area of the West coast of South America during the period 1150–1532. The Spanish conquered them from 1532 onwards. They spoke a language called Quechua which is still spoken in the Andes today.

4 The land and sea as a route to trading and spreading knowledge.
Vessels were trading on the river Nile and Euphrates in 4,000 BC, and probably much before that time. The Phoenicians were great sailors and dominated the Mediterranean around the period 1500 BC, together with Mycenaeans of Greece. The Minoans of Crete became powerful traders and developed great wealth, resulting in a big navy by 1700 BC. The seafarers and their sailing ships were sufficiently advanced to cope with open sea voyages. In 285 BC a lighthouse was erected at Alexandra over 300 feet high that could be seen 25 miles away and guide ships into port. The market trading area was close to the beach and harbour. The pharos lighthouse was destroyed by an earthquake in 1375. The art of paper making had been brought from China 2,000 BC, so ship captains could meet

in port to swap details of trading routes and harbours, to be recorded in the form of charts. The first cities were built in Mesopotamia in about 7,000 BC. With this degree of skill, carpenters must have been able to build sea going vessels capable of open sea voyages. Certainly 2,000 years ago trade routes were established with Northern Europe and the tin trade with Cornwall. The Book of Enoch, one of the books once part of the Bible describes Henges in the UK and how to set them up as observatories so there is evidence for trade routes even earlier. The Phoenician traders were travelling all along the British coast. Ivory was as precious as gold and Walrus ivory could only be obtained from northern latitudes as the creatures were gradually hunted out further south. If you can sail 30 miles on the open sea, with a bit of luck and good weather, you can sail 3,000 miles. The trade winds blow from the Mediterranean to the Caribbean at certain times of the year. With the wind behind, it isn't difficult to sail the Atlantic. I have done it several times in my small sailing yacht. I have the advantage of a chart and sextant, so when I arrive in the Caribbean I know my position and the land that lies ahead. So I am certain sailing vessels crossed the Atlantic 3,000 BC. Could other sailors have come across the pacific on the prevailing westerly winds by taking the north route? They could have then sailed down the west coast of South America to reach Peru and the sites of the beginning of time at Lake Titicaca. There are accounts of white skin man with red hair (Caucasian) arriving by boat without oars. In Mexico City the legends of Quetzacoatl and his immense knowledge and understanding of all things in ancient times, spread far and wide. Quetzacoatl taught the measure of time and astronomy. Time would enable positions at sea to be found. He was referred to as a God King of the Toltecs. There is a record of him leaving Mexico for the Mayan Lands of Yucatan in AD 986. Man had crossed from Siberia to Alaska 20,000 years ago, and tribes had moved south into America. It was only a question of time before they moved into South America, and the River Amazon became populated. Where did the designers of the ancient temples get their Egyptian similarities? Could an Egyptian have landed in South America?

Indian d'hows were known to make sea voyages as long ago as 3,000 BC. These voyages were to the Persian Gulf to collect copper and trade with crops and ornaments. They sailed 1,000 miles on the open sea and although much of it was coastal hopping, the vessels had to be seaworthy for such a voyage. These types of vessels may have crossed the Atlantic as long ago as 3,000 BC and provided the influence of Egyptian architecture seen in South America. Trading routes to India from Persia were well known by

4,000 BC. In 2,000 BC trading routes to China were established to bring back silks to the royal courts.

You would not have expected people from as far away as China and Japan, Egypt and America to have shared secrets and customs at this time. Did their knowledge come from heaven in fiery chariots which is described in the Bava Ghita? (Ancient Hindu texts).

5 The Influence of Religion

The gods lived in the sky. The souls of the dead passed to the gods to live in immortality. The gods sent thunder, lightening, rainbows. It was necessary to please the gods. What other beliefs could Homo sapiens have? He did not have a scientific training that could explain an alternative view. His obsession with death must have imprinted some strong memory patterns on his brain. Brain cells may be allocated a space for religious beliefs. Religion or beliefs can be forced on you, a bit like politicians who indicate if you do not vote for me you will not have a job. How dangerous to build on other ideas from a wrong assumption.

James Ussher had been chaplain. to King Charles 1. His life's work, a study of ancient history, from the day of creation, 'the annals of the world' was published in 1650. What can we learn from him? Build on a false statement and getting the backing of the church, people will believe anything. Based on the bible, the time of Jesus, and who begat who in the Old Testament, Ussher said 'at 6.00 p.m. on 22 October 4004 BC God created the world. This was a precise date for the age of the universe. It was even printed in bibles in 1900 alongside the opening verse of Genesis. Ussher was an educated and learned scholar. Why was his thoughts on the date of creation allowed to misguide the public so long? As a protestant he came under criticism from the Roman Catholic Church, that the whole doctrine of Protestantism was false. His scholarly works dispelled this idea, and few were prepared to challenge his knowledge. The Irish rebellion of 1641 left Ussher penniless with civil war raging on his doorstep. Despite this with his superior knowledge, he persisted in his claim. These claims would have died a natural death had it not been for the University of Oxford publishing bibles under the contract entered into with Thomas Guy who printed an illustrated bible with the date of creation of the universe according to Ussher. Together with Thomas Guy's other business investments the sale of bibles made good money, which he donated to make Guy's Hospital in London.

Now the date of creation was fixed, it was up to others to disprove it.

Ancient China offered to predate Ussher's claims. Fossils were discovered suggesting the remains of once living animals, and suggesting a much earlier date for creation. Knowledge is a step by step process, and the influence of the church often opposed the findings. Galileo 1564–1642 made telescopes, microscopes, thermometers. He made great advancements in mathematics and astronomy. He was put in prison for making his finds public. With the advancement of science came the industrial revolution, There was the great exhibition in London in 1851, Darwin's theory of evolution, origin of the species, was still being questioned in 1904 and the church was still holding onto Ussher's time of creation. If the church is going to act so stupidly and dogmatically, is it not time we question their motives?

Conclusions

Egyptians were great at geometry, which leads to engineering. It is hardly surprising their influence would eventually be felt worldwide. The precise aligning of their massive structures to the geographical axis of the earth, to the East and West and the detailed knowledge of precession of the earth's wobble through the stars of 1° every 72 years, returning to the original point of first wobble in 26,000 years, leaves us wondering when this knowledge was first discovered and how far the knowledge had spread. Where did their religious knowledge come from? There were libraries containing details of early history sadly destroyed at the time of the Spanish inquisition. The legends tell us of men that came down from heaven. We need to find out by excavating of ancient sites in Egypt and America, which could be 30,000 years old, to discover our past and try to date the findings more accurately, so more pieces of the jigsaw puzzle can be assembled. Can we find the 42 books of Troth and decipher the ancient hieroglyphics. What messages have been given to us by the ancients hidden away in temples and pyramids?

Here are some of their messages:

1 Ancient texts were written recording the early history of mankind. The importance of these texts was recorded by ancient philosophers and scribes who saw the texts. They have been lost. Khufu searched for the texts in 2500 BC, and Rameses 1250 BC claimed to have found one book of Troth.

2 The ancients believed in the importance of bringing down the sky to the earth as everything is made by the gods in the sky – the Nasca lines indicate this meaning.

3 That the brain responds to stimulation to give an appearance called reality. The understanding of knowledge can allow the higher part of the brain to pass the soul on into immortality amongst the stars.

4 Religion teaches us a code of practice. The ancients wanted to pass a message to us. The need to be peaceful.

5 Great teachers came to earth, claiming to have all knowledge.

6 The Egyptians and others developed great geometry skills a long time ago, measuring the rising and setting of the sun, moon, planets and stars. Each country appears to have developed independently from one another, whether in Egypt, China, Mexico and South America. Egyptian influences can be seen in the construction of temples.

Books Recommended for Further Reading

1 The Hieroglyphs Handbook by Philip Ardagh – Faber, Faber Ltd, 3 Queen Square, London WC1N 3AU

2 The Atlas of the Ancient World by Margaret Oliphant – Marshall Publishing, 170 Piccadilly, London W1V 9DD

3 The Encyclopaedia of the Ancient Americas by Jen Green, Fiona MacDonald, Philip Steele, Michael Stotter – Anness Publishing Ltd, Hermes House, 88–89 Blackfriars Road, London SE 18 8HA

4 Mysteries of the Ancient World – Geoglyphs by Paul Bahn, Orion Publishing

5 Mysteries of the Ancient World – Decoding the Stones by Steven Snape, Orion Publishing

6 Mysteries of the Ancient World – Mummies Unwrapping The Past by Rosalie David, Orion Publishing.

7 Mysteries of the Ancient World – In Search of King Minos by Louise Steel, Orion Publishing

8 Mysteries of the Ancient World – The Pyramids Star Chambers by Robert Bauval, Orion Publishing

9 Mysteries of the Ancient World – The Easter Island Enigma, Paul G Bahn, Orion Publishing

Chapter 5

The brain

EVOLUTION AND INHERITANCE

The current theory why the brain has developed so large in humans, as opposed to chimpanzees, when our brains were similar five million years ago is attributed to various factors, communication being important. Living on the ground we had to defend ourselves against aggressors, learn the art of making spears and sharp tools for cutting, to build shelters for our families, learn to use our hands, and hunting skills. This required the need to communicate, to work in groups, to develop a language, learning, passing on our skills to our children and eventually the written language. This increased our neuron connections, the computing powers of our brain, the ability to plan ahead and our reasoning powers. Slowly we have become masters of planet earth, controlling animals. The climate played a major part in survival of the fittest, either from cold or from drought. Life was nomadic, on the move looking for food, water and shelter.

Five million years ago man and chimp were almost identical. Today a baby chimpanzee displays the same characteristics as a baby human

What caused the difference in chimp and human brains? The only clues I can think of are 1) The chimp learnt to climb trees to escape dangers. He developed simple language but never developed a specialised speech neuronal circuitry. 2) Man on the ground needed to make spears and sharp tools to defend himself, to learn how to use his hands, and the need to communicate by speech. Man's primitive life style did not require a bigger brain to operate these functions. His communication was by sign language, probably making shapes in the sand with a stick. When language was invented and the written word added to communication, then ideas could be collectively recorded and passed on in a learning process, to be able to recall in the memory. I can understand this stimulation may require the brain as a computer to expand. The written word had not been invented, and yet we find a fully formed large brain in man 125,000 years ago.

The enlargement of the human brain is associated with the development of the cortex and prefrontal lobes, and correspondingly larger skull sizes. The skulls of man, one million years ago, Homo erectus, showed little signs of this development. Some major stimulation occurred to the developing brain between one million years ago and 150,000 years ago.

What triggered this development? I can understand a system when not enough information is fed into it for it to move forward. If the same system received new information from several different sources, it could expand at a varying rate. It is as if the process of knowledge is a slow build up of facts which can accelerate or retard the overall development. In nature we see systems that go wrong. In meteorology, when too many variables are fed into the equations, we see a state of chaos, only to return to predictability when some variables are removed, and a weather report can be given. In the acceleration of knowledge, did the brain display an element of uncontrolled growth?

Is the final development of our brain best suited to our present environment, and can the brain be altered to accommodate any changes? Why is the brain is so large? We must be looking at one of two possibilities. Either we have inherited our brain size from a superior race 500,000 years ago or for some unexplained reason the brain has decided to grow so big. There is not the stimulation for a large brain to develop from small groups of people leading a nomadic life with no written language, and unbelievable hardships, leading lives like animals, with a low expectancy of life.

Conclusions on How We Have Developed

Five million years ago chimpanzees climbed trees to escape predators but did not develop a specialised neural connection for speech that we have inherited today.

Apes grew so big and muscular they had a problem climbing trees, but presented a fearsome sight for would be aggressors.

Walking on two legs. Footprints found in Africa have been dated at three million years old. These primates and humans are collectively called Huminids. There were several types between one and four million years ago. A skeleton found in South Africa has been dated at three million years ago, their brain sizes only a quarter of our brain size today. The name Austracopithecus is given to this ape which walked upright two million years ago. Homo Erectus is the name given to the primitive man between one and a half million years ago. He had a small brain and lived in Africa and South East Asia.

Why did we decide to walk on two legs? It gave us hands tree to pick the fruits off the trees. This did not provide the stimulus to increase our brain size.

Homo-sapiens, 250,000 years ago. This is the date given for the arrival of man with a bigger brain. His name is Neanderthal man and he would have bred with our ancestors. He roamed Africa, Europe and Asia, and had to cope with ice ages and world disasters. His appearance was short and stocky and a large nose. Perhaps this helped in some way with coping with the cold. He may have wandered south to escape the cold. Another genetic line was migrating north from Africa about 100,000 years ago. These were better adapted to survival and dominated the Neanderthals. Slowly they replaced Neanderthal man and there is no evidence of genetic survival of Neanderthals in modern man. If you compare the DNA of Europeans, Africans and Australians today, the variations are so little it suggests our African origin is most likely.

Many different races appear to have entered America including genes from China. The ice age in 20,000 BC came down as low as North Florida, which must have been terrible for the roaming nomads, and I expect they had to move into South America.

Man's brain appears to have been well-developed 100,000 years ago. There is a big difference in stating the brain was fully formed, and stating it was fully used. Unless the brain is used, the density of neural pathways is not made. Einstein did not have a bigger brain. He had more dense neural connections.

Evolution of Skin Colour

Sunlight, ultra-violet light makes vitamin D in our skin. When we move north, skin colour goes lighter because of the less effect of ultra-violet on the skin. Too much sunlight, and we can get skin cancers, melanomas. This is the reason why we need sun protection cream when we go on holiday to the sunshine.

We have inherited a large brain. Now we must try to find out how it works, and how to use it to best advantage. Is it really only a complex computer?

Structure

Our brain consists of billions of circuits connecting all areas, a complex of inherited patterns and evolutionary changes. From childhood times we record our surroundings and develop our sense of sight, recognising our mother's face within the first year. We develop our senses of recognition, hearing, smell, taste, balance, our first words leading to speech, communication, language and use of our hands. We can follow the development of a child's brain. We have an insatiable appetite to use our brains to discover more about our surroundings. There are levels of recognition in the brain to direct our attentions where we want and blot out less relevant things. The stored memory of all objects are recalled by sight memory to give conscious recognition. One pathway interacts with the outside and a different pathway recognises what things are and turns seeing into understanding.

The Human Brain

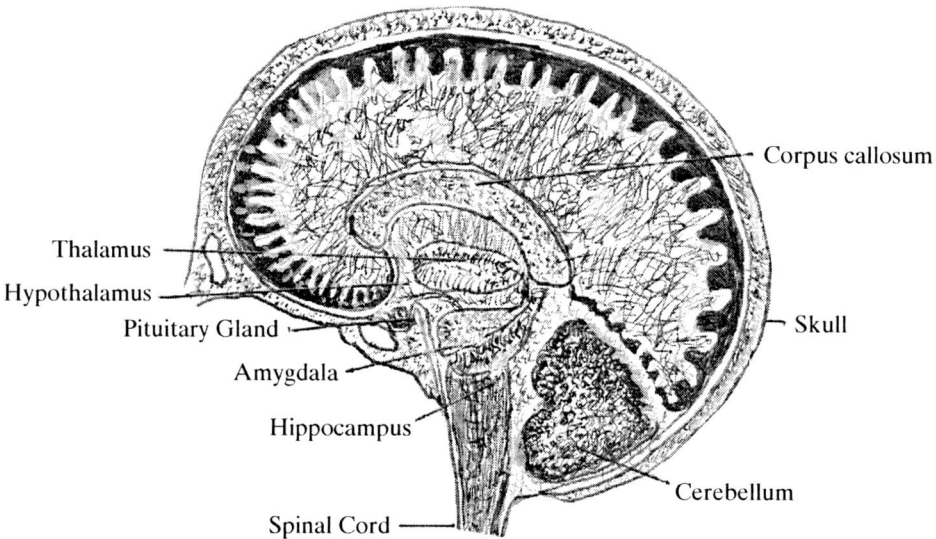

All mammals share the three brain regions. The cortex in humans is very large and to fit into the skull the cortex has to fold up, forming distinct lobes, separated by deep fissures, giving a wrinkly appearance like a walnut. As a child's brain develops, the soft calcium of the skull expands to accommodate the brain's enlargement until the pituitary gland secretes a hormone which stimulates the ossification of the skull; bone is laid down to strengthen the skull, so it can no longer expand. The cerebrum, the uppermost part of the brain, has two cerebral hemispheres, like two brains joined together to produce one response.

The density of neuron connections adds to the overall computing power of the brain, We have more neuron connections in the visual cortex. When a teacher writes on the blackboard and explains things, we take down notes and write them up at home in a neat copy, and add to them from reference books. When we go back to read and learn our notes, the visual cortex lays down memory patterns for recalling at will. If we had recorded the lecture on a tape recorder and played it back hoping to learn from it, the result would be disappointing. The hearing cortex comes into its own for musicians who can 'hear' the music in their brain. We develop our brains by studying.

Different parts of the brain have different special functions. We can trace the optical nerves from the eyes to the visual cortex. Vision is one of our oldest senses. The eye started to form 250 million years ago. Nerve cells are called neurons, and form a web of interconnected networks. Under the scanning microscope we can see their structure. There are over one hundred billion neurons and also many other cells. There are ten times as many glial cells as neurons. They wander about in the brain repairing damage and collecting debris. The neuron has thousands of tiny filaments, or branches called dendrites which receive the incoming signal to the cell nucleus. This triggers a response from the cell nucleus which goes out by a different route called an axon. Each neuron has one axon. It is a constant diameter but varies in length. At the end of an axon, there are a number of small branches which pass on an electrical signal via synapses to connecting cells. So neurons fire electrical signals and chemistry takes place. Sodium and potassium ions interchange through the cell membrane and a potential difference causes electricity to flow across a synapse. Neurotransmitters are chemicals affecting this transmission. A protein called nerve growth factor (NGF) helps axons and dendrites to grow, and has other effects.

The human brain can make new connections readily. The stimulus is our conscious ability to learn and refine our skills. This answers my earlier question; 'can the brain be altered to accommodate our new environment'? The answer is yes. To answer another question which I asked. Is the brain like a complex computer? The answer is yes and no, basically no, because computers learn by being programmed in advance. The brain learns by altering its internal circuitry. We need to build a brain to model its essential functions. Unfortunately we will also need to build a body to provide the incoming signals. This makes the complexity of the task too great. If we cannot define consciousness how do we make it? How do we build in emotions and feelings?

The Cerebellum

The cerebellum collects incoming signals from the spinal cord and sensory endings and co-ordinates movement. In damage it produces clumsy movements. The cerebellum will co-ordinate the input from the senses of piano playing and driving a car which I will describe in detail in this chapter.

The Cerebrospinal Fluid

The brain and spinal chord are surrounded by fluid which can be analysed in disease by a lumbar puncture. The fluid is crystal clear and contains cells, protein, chlorides and glucose. Bacteria can cause meningitis and examination of the cerebrospinal fluid shows an excess of cells in this dangerous condition.

The Cortex

These deeply folded layers of the uppermost part of the brain give the appearance of a walnut. We display a large cortex. It is as if the brain wanted to expand further but there was not room. The cortex is regarded as the site of higher functions of reasoning, and possibly emotions.

The Prefrontal Cortex

The prefrontal cortex is responsible for the 'working memory' amongst other functions. It co-ordinates the information, together with other parts of the brain, to produce the necessary data to operate in the current situation. It is closely linked to emotion, feeling and thought. In man it is relatively much bigger than in chimpanzees or cats, and accounts for 29% of brain volume in humans. It is associated with higher processes. Damage to the prefrontal cortex leads to a change in personality, the difficulty to operate the working memory. The psychologist Larry Squire, has a good definition of working memory 'placement of remembered events in their proper context'. The prefrontal cortex allows us to plan ahead. Chimpanzees can only plan ahead a short time. The prefrontal cortex together with the hypothalamus can act as a time clock for body functions.

I wonder if the prefrontal cortex plays a part in imagination and dreaming. Brain scans of people suffering from depression show an overactive prefrontal cortex.

Leucotomy

People displaying serious aggression can be controlled by surgical incisions or removal of part of the prefrontal cortex

The Hippocampus
There is evidence to suggest this area is associated with the long term memory of events. Scans revealing damage to the Hippocampus are associated with memory loss. The memory of facts is attributed to another area, the temporal cortex. Experiments on birds show the Hippocampus enlarges with teaching. In rats removal of the Hippocampus produces a loss in its sense of direction. Emotional stress can damage the Hippocampus, with less ability to lay down new memory patterns. Research suggests this is true in animals.

Amygdala
Damage to this area leads to indiscriminate sexual behaviour. The feel of an emotion is linked to the cortex. The connections of the Amygdala to the cortex send stronger signals i.e. emotions dominate our thinking, but thinking doesn't dominate emotions.

Thalamus or Two Thalami
This is associated with laying memory patterns down.

The Hypothalamus
The hypothalamus controls hormones e.g. hunger, sex, growth and the immune system. It can trigger arousal.

The Pituitary Gland
It produces hormones affecting growth, somatotrophic hormone STH, thyroid stimulating hormone and other bodily functions.

We do not know what all the areas of the brain are for. If man has been so obsessed with immortality, how to live forever, perhaps an area of the brain has developed to receive these signals of religion, early beliefs. The prefrontal lobe areas of our brain are associated with behavioural patterns, sociability, and the ability to think through a problem, to plan ahead. This area may be the most sophisticated and recently developed.

How can we find out more about the brain?
a)   There are chemicals, hormones, neurotransmitters affecting the brain and the effect of drugs to consider.
b)   A study of damage to the brain by injury, strokes, diseases, doctor's cases.
c)   Brain scanning methods (including the developing baby's brain).
d)   Surgery.

e) Normality and sub-normality.
f) Hypnosis.
g) Processes of learning and laying down memory patterns for recalling. The baby's brain.
h) Experiments and studies on animal brains?
i) Genetics.
j) Conclusions.

(a) Chemicals Affecting the Brain are:
1 Neurotransmitters
2 Hormones acting much slower
3 Drugs taken for special effect on the brain, e.g. painkillers, anaesthetics, hallucinogenics etc.
4 General body release in response to emotions, e.g. fear releases adrenaline – peptides.
5 Chemicals directly related to control of emotion and higher levels of behaviour pattern, e.g. serotonin.
6 Building blocks for nerve cells and semi-permeable membrane, e.g. sodium, potassium, calcium.

There are many neurotransmitters in the brain, built up from amino acids. Science today is constantly finding new neurotransmitters. When their purpose is known and the effects of too little or too much are noted, it is the next step to be able to control them in treating a medical condition.

How Food Can Play a Part
A list of vitamins is included in chapter 10(b)

1 Bananas are rich in potassium, needed for brain cells. This is true of coffee in small doses.
2 Nuts contain magnesium and amino acid. Tyrosine.
3 Vitamin C is involved in noradrenaline production – fruit.
4 Pasta – this is a source of the amino acid tryptophan which makes serotin.
5 Cereal – this is a source vitamin B and calcium.
6 Chocolate – a source of iron – milk chocolate contains calcium.
7 Serotonin – Controls emotions – low levels as a result of strict diets can lead to unstable moods.

100

8 Endorphins – control pain very well and transmission of nerve impulses.

9 Pangamic Acid – B15. This oxygenates the tissues and slows the ageing process. The Russians are very enthusiastic about B15.

Some food additives may be harmful to us, not only affecting neurotransmitters but acting as carcinogenic agents. More research is needed into food additives.

(b) Cases Doctors Tell Us About Where Brain Damage Has Occurred

Case 1
A man had a stroke. A brain scan revealed the pre-frontal lobes were damaged. The man could not think through the outcome of his own actions.

Case 2
A child was found to be having continuous epileptic attacks. A brain scan revealed massive damage to one side of the brain. This side of the brain about a half of the brain was removed, and the good side took over and the young brain took on some of the functions lost by the other half.

Case 3
A man displays the ability to draw different pictures with his left and right hands at the same time. The brain in this case appears to be operating separately in its two halves

Case 4
A lady friend of mine complained 'I keep dropping things and sometimes fall over. What's wrong with me? I've been to my doctor and he says give up drinking gin'. The dilated pupils, general muscular wasting indicated something seriously wrong. A brain scan at Southampton General Hospital showed degeneration of the cerebellum. The brain cells were dying in a basal part of the brain which appears to be a sorting house for basic bodily inputs. Were these nerve cells not getting their nutrients? Did the ageing process go wrong, the body time clock say 'you are now growing old, your time is up?' It was an inherited condition. Genetics played a part. Her mother died at exactly the same age of 61, of the same condition.

Case 5
I have seen a child in a sub-normal hospital with no arms or legs, who cannot see or hear. The brain was considered to be working normally. A lady would arrive and cuddle the child with words my little darling. The stumps would waggle in apparent excitement. The child's brain would never develop properly with no input into the memory bank, and yet the brain indicated there was consciousness.

Case 6
A man with a stroke to the right side of the brain is unable to complete a picture on the left side. In his mind drawing the picture may be complete, but part is missing. The brain tells him the picture is complete when it is not.

Case 7
A blind person. There are sensory feelings in the fingers and similar sensory feelings in the visual cortex, blind people can feel brail and visualize the feeling

Case 8
Take the case of a person who has lost a hand. Perhaps the stroking of the stump will produce feelings of finger movements in a 'phantom hand'. After a period of time, the brain may move its connections and stroking a part of the face produces a feeling that the fingers are moving. 'Hey Mr Brain'.[1] I can visually see there is no hand so there is no need to pretend. Re-allocated those useful brain cells to another function. No, the brain cannot.

(c) Brain Scanning Methods
1 The early investigations consisted of placing electrodes on the scalp and measuring brain waves against a conscious decision.
2 Standard photography.
3 X-ray computed tomography.
4 Positron emission tomography.
5 Nuclear magnetic tomography.
6 Voltammetry detects currents and enables us to detect the release of neurotransmitters.

Abbreviations used in this line of diagnosis and research:

MEG = Magneto encephalography.
PET = Position emission tomography.

EEG = Electroencephalograms.
FMRI = Functional magnetic resonance imaging.
ECT = Electro convulsive therapy.
LTP = Long term potentiation.

MEG scanning is a more advanced version of EEG. It can detect the smallest areas of electrical activity produced in the cortex.

Conclusions from Imaging
No one part of the brain is exclusively active in any one mental effort.

(d) Surgery
A lot can be learnt about brain function during surgery. To avoid damaging the brain in surgery the patient is made conscious to assist in the operation. Placing two fine electrodes on the brain can produce sensations indicating the function of an area. The patient counts to 10. Passing a micro-current stops the patient from completing the count. A lot can be learnt about brain function during operations.

(e) What Conditions Give Rise to Sub-normality? What is Normality?
Normality is an acceptable response from a given stimulus. Sub-normality was once described as an IQ below 65. It suggests that the brain is not functioning normally at some point in its mechanism. There are levels of sub-normality often associated with worsening degrees of social behaviour and ability to look after oneself. Do genetics transmit sub-normality? Yes they do. I came across an interesting case working in a subnormal hospital.

It was a surprise at the subnormal hospital to find patients displaying large protrusive jaws, heavily ossified bones with narrow skulls, too many cases looking the same. The conclusion I came to is that the anterior lobe of the pituitary gland had secreted a hormone to start ossification too early, and the skulls became ossified before the brain had a chance to reach its growth. This could have accounted for their sub-normality.

If nature can make a mistake that way, why can we not have later ossification and bigger brains.

(f) Hypnosis
In hypnosis I can offer you alternatives; these are reasonable propositions. I will give you an example of what I am trying to do 'to bypass your critical factor'.

Close your eyes, outstretch your arm and imagine a blackboard ahead of you that you are going to write on. Take some white chalk and starting at the top of the blackboard 99. Think clearly in you mind. Can you see what you have written, 99? Now write 98 underneath it, and 97 beneath it, decreasing numbers, write 96, 95, 94, 93, 92, 91, 90, 89 etc. One number under the next. You will find that with your arm outstretched to the blackboard, writing all the numbers will make you arm feel tired and it is falling. Let it fall down gently onto your thighs, helping you to feel relaxed and tired. Close your eyelids very tight and feel the tiredness creeping into your eyes, making you feel sleepy. Other suggestions follow and you obey the commands of the hypnotist.

Let us analyse what has happened. The hypnotist says your arm is tired and falling as a result of writing on the blackboard, with your arm outstretched. This is reasonable. The brain accepted it. The reason why your arm is falling is you are writing one number under another. An analytical mind may have analysed every word and sentence, and asked why? Making the hypnotist's job more difficult. Young children are easy to hypnotise. They are more open to imagination and suggestion.

Now you have your subject under hypnosis try regression. We are going back in time to when you are a child. You are only 5 years old. I am going to give you a pencil and paper so you can write your name. Perhaps the legs will splay and the arms outstretch giving maximum body support as the child takes the pen and writes at the age of 5. Mum confirms the signature with one from that age. What can we learn form this? The brain has recorded not only the writing at the age 5, but the body muscles and controls that produced that writing.

Hypnotists can work with big audiences. For example, I want everyone to place their hands on top of their head and lock the fingers together. Imagine your hands are in a vice. They are locked together. You cannot separate them. Will everyone who cannot unlock his or her hands please come up on to the stage. The hypnotist may say 'there's a big black dog'. They look behind. It is there. The hypnotist will bring you back into the real world, cancelling any suggestions. The big, black dog has now gone. I want you to imagine you have had a good sleep. I am going to count down from 10–1 and you will be totally relaxed and wide awake, or some such words.

What would happen if the hypnotist left you hypnotised. You would walk around, maybe for 3 days, looking behind and seeing a big, black dog, until eventually the suggestion would wear off

On one occasion, I was hypnotising a man at a party. The lights were out, the man lay on a couch and I said 'imagine you are on a beach, a sandy beach. It is hot and sunny. Your whole body feels warm. The tide is coming in and will soon reach your feet. The sea is only 2 feet from your feet. The next wave is going to cover your feet. It is going to feel cold. Here comes the breaking wave, yes its going to cover your feet now. Your feet feel cold'. Just at that moment someone came into the room, put on the lights and said 'what's all this nonsense going on? The party broke up, connection was lost. The man left, and walked about for 3 days with cold feet until the feeling wore off. It is imperative you cancel the suggestion. It can be dangerous when the brain is very confused and reacts badly. I will give you examples of these but first let's look at the remarkable time factor.

Under hypnosis, you can try this one.

'I am going to put your bracelet in the back of the a drawer under a book. You will not be able to find it. In 21 days time at 7.00 p.m. on that evening, you will go to the drawer and find your bracelet. So it happens, the brain's time clock'.

Your brain is susceptible to suggestion. Learn to analyse what is being said to you, listen carefully and ask yourself 'What does this mean?' A salesman might be trying to trick you into buying something you don't want. The brain has to receive information, sort it, analyse it, lay down memory patterns for recalling, recall it, reproduce it and transmit the message. The brain is an enormous storage tank, all in memory patterns. The whole brain is used in this storage mechanism. Is this storage capacity unlimited? No. I found during 6 years of a full time University course trying to remember more and more facts and at the end of 4 years was becoming progressively harder. I am only speaking for myself.

Where there are no sensory endings the brain cannot necessarily respond. If I hypnotise you, I can pinch a piece of skin together and say, 'this area I am marking out is going numb; I am going to put a needle through it and you will not feel it, and when I withdraw the needle it will not bleed.' 'Hey Mr Brain.' You can control an axon reflex to stop it bleeding. Perhaps if I hypnotised you and told you to stop the blood supply to the malignant tumour cells on your brain you would do it? What sort of a supreme brain power do you have that you allow something to attach itself to you and kill you? Do I have to draw up a list of your limitations? 'I do not have sensory endings on my brain surface to do this!'

Have we learnt anything about the workings of the brain?

1   The brain has an input factor, interpretation and response.

2   Regression shows memory may be stored in patterns – involving other conditions present at that time.
3   There are different levels of strengths for storage and recalling.
4   The brain has a built-in clock.
5   Spacing – we have not talked about this – it gives the interval for comparison.
6   Intelligence – one brain is more intelligent than another.
7   Evolution – takes a long time to change things, genetics are transmitted continuously.
8   Genes and biochemistry control most things. It is not unreasonable to consider that genes control:
   8.1 Intelligence
   8.2 The brain's time machine.
   8.3 The brain's time spacing.

An example of spacing is perfect pitch. One in 10,000 people have it. Everyone can develop it. Birds have it, so they can sing a perfectly pitched note and other birds can receive it. If you have it, you can recognise any note of 88 notes on a piano and name it. Perfect pitch is a point in the space of hearing. From this position other positions can be given appropriate values in space. You can hear the pitch of a note in your brain but you cannot necessarily sing it. It isn't always possible to reproduce on the response side, the response you may want from your brain.

   When you get older, perfect pitch goes up a semi-tone which is confusing for a piano player listening to music. Nobody knows why. The hearing memory is a very important part of the pianist's learning. The brain gauges spacing again in a different way for the pianist. In piano playing the space between notes is accurately recorded and allows you to jump big distances on the piano and with practice always hit the right note without time to see it. Another example is touch spacing. This could be described as finger pressure sensitivity producing an effect validated by hearing. Again, it is a graduated response about a recognised level. In one chord of notes, different notes can be accentuated by different fingers. The ability of the brain to think and send message so fast to all different fingers is extraordinary, when the next chord may be different. The concert pianist will remember an entire piece, a piano concerto, without music. A concert pianist may have converted to memory 80 piano concertos, a quite remarkable example of memory, e.g. the late John Ogden.

(g) Let's Analyse Some Examples of How the Brain Learns
1   Walking – the automatic response.
2   Learning to drive a car.
3   Learning to play the church organ, with hand and foot movements all going in different directions.
4   The concert pianist sight-reading 1000 notes on a page of music he has never seen before.
5   The developing baby's brain.

1 Walking – the automatic response
Man and animals walking on two legs develop the fine art of balancing on the soles of our feet. This all stems in childhood from crawling to strengthening our leg muscles, balancing, standing upright and walking. This goes into an automatic response. When we walk we are still reminded of our surroundings. Eyes can tell us what we see under our feet, memory can tell us what to expect and what to avoid. The sole of the foot can be given a feeling of the ground ahead before we tread on it, whether it is hard ground or soft ground, slippery and wet, firm and dry, hot or cold, sharp and uneven to walk on, and many other factors. The brain does this automatically so we do not have to stop to think about it. We may have to give greater analysis to one aspect or another, e.g. am I going to sink in the soft ground. As a child, curiosity is a motivating force to find out. We have to learn and often learn by making mistakes. As we walk, so we may decide to go faster and run. Brain responses have to be quicker to cope. As we walk, the sole of the foot acts as a pump in the venous return of the blood to the heart.

In all these examples I am over simplifying a vast amount of action taking place.

2 Learning to drive a car
Because many of us learn to drive a car, it is an experience we can all associate with. The brain starts recording and analysing from the moment you open the car door.

1   Getting into the car, the driver seat can you see out of the window, is the seat high enough, comfortable (probably the last thing you think of at this moment)?
2   Can you reach the pedals? What foot controls are there? How do they work? How far down do you have to press them?

3   Let's hold the steering wheel – how does it feel? How wide is it? Can we grip it.?

4   Where is the brake, the footbrake and the handbrake? The brain is recording how to operate the brakes.

5   The brain has recorded many things. You can hold the steering wheel, look out of the window, put your foot on the brake to stop, and other pedals are clutch and accelerator.

6   Now more complex learning – the brain has to cope with so many different things at once – the gear lever, clutch control, accelerator, speeds in different gears, changing gear – practice make perfect.

7   At first this all seems an improbable task, except we can see other people driving cars, so it must be possible, and we all have confidence in our teacher – the driving instructor.

8   Our first drive – apprehension – body and mind muster all alarm signals – will it work? Start the engine, hands on wheel into first gear, handbrake off, look in the mirror, signal, off with the clutch, and steer slowly at first, bumpty, bumpty, and the sudden jerking movement is resolved when the engine stops (stalls).

9   Every sense is coming into play, group patterns of learning are getting so familiar they are going into automatic response. After many weeks of learning we don't need to mentally think about how to drive. It has become an automatic response. Now we need to give our attention to all round vision, awareness of our moving surrounds, judging distance in approaching, overtaking, stopping, turning, the need to exercise care to avoid a crash. Our examiner knows the level of skills needed to pass his test. We have created a standard of knowledge the brain has to achieve.

3 Playing the church organ

We have many things happening at once. We may have more than one keyboard, console, organs stops to pull out, both hands often moving in contrary directions and to confuse matters foot pedals to play notes. If we had 3 brains all joined together, one part could control the left hand, another part the right hand, and the third part the foot movements. Then a certain area in the brain could control all three centres. The brain doesn't work like that. It arrives at its ability to cope with this seemingly impossible task in a different way. Similarly to learning the piano, each hand will be practised separately, bar by bar, phrase by phrase, and in small sections both hands put together, slowly at first, building to the required speed. The foot movements need to be practised and then all three put together. Eventually the brain recognises the piece so well, some information goes into automatic response, giving more attention to other parts of the piece. We are talking about patterns of familiarity and comparing them to known brain pattern, using finger memory and hearing.

4 Playing the piano

How can a concert pianist sight read music of a complex nature, e.g. play music from a sheet of music that he has never seen before?

Each stage of piano learning goes into automatic response, it takes an enormous amount of practice to be able to recognise groups of notes on a page of music and play these notes on the piano. An easy starting point is to play one note at a time, a simple tune, one hand only. The 88 notes on a piano have to be identified, with a reference point in the middle of the piano called 'middle C'. A student is taught an octave of 8 notes, which can be repeated up and down the piano. Vision plays a part; fingering is used best suited to play the notes with ease and evenness; hearing plays an important part in musical memory.

The notes are not of equal tone. The interval between 3, 4 and 7, 8 notes in the octave are only half the value of the value of other notes, e.g. halftone.

The brain is learning spacing, position of all notes on the piano, the leaps of the hand to get to all the notes and what the individual notes sound like to the ear, e.g. recognising individual notes = pitch.

So far the brain has learnt the name of the notes, how they are put on to paper to read, the 88 notes of the piano arranged in octaves with full tones and semi-tones, to identify middle C and similar C's up and down the piano, giving us distance and spacing with black and white notes. The memory is recording this in vision and audio areas of the brain. The fingers now develop a degree of touch sensitivity according to how they depress the keys. We have not yet introduced the sustaining pedal operated by one's foot.

We have studied the piano, the mechanics of how it works, playing with one hand at a time, familiarise ourselves with notes, octaves, intervals. But like driving the car, it is not simple.

We must learn with both hands and put them together, learning not only by looking at the sheet music, but listening to the sound produced to arrive at our first experience of 'finger memory'. So finger memory is a slow build up of the visual notes on the paper corresponding to notes on the keyboard and hearing memory. Playing more than one note at a time leads to chords, which are more pleasing and we can hear the harmonics ring by the soundboard. The melody might rise and fall, the memory recognising this rising and falling, and the way notes can be phrased together in a tune. The harmonies produced introduce the pleasing sound the hearing memory can understand. Soft passages may alternative with loud passages. The way notes are grouped together, phrasing is indicated in the written music, and also the time signature. Rhythm can dominate a piece of music or be constantly changing. Black notes can produce more semi-tones and octaves can be played with key signatures starting in sharps and flats. Music written has recognisable characteristics e.g. key signatures, timing, groups of notes, chords. The brain will eventually learn to recognise all these things and a lot more. In one cord you may have five different notes to play at once. The extraordinary thing is the brain can tell the fingers to play 5 notes at once with different emphasis of touch on each finger. The brain will come to recognise from memory at least 40 different automatic responses making the sight-reading of complex music possible. A composer may indicate how to play his music, e.g. Chopin, or Strauss. What a satisfying achievement in life to reach the top of the ladder as a concert pianist. A detailed analysis of teaching the piano produces grades to be achieved. This leads to examination at the stages of these grades. This produces a predictable pattern of learning at which we can suggest levels that can be achieved in a given time.

Why can one person learn to play so much better and quicker than another? I have heard a sub-normal patient play the piano. The connections to the music centre of the brain were perfectly intact with the ability to activate the necessary muscular controls and all relevant senses and memory patterns. I would suggest some students of piano display better brain power at playing, e.g. brain connections to relevant areas are more reinforced, and together with good piano hands naturally find the process easier. Is genius not one stage better than that? Very few piano students become concert pianists. Often great musicians come from musical families, so learning at an early age is of great benefit; look at the Bach family – 56 brilliant musicians. Franz Liszt was a genius, both as a concert pianist by the age of 16 and as an innovator of piano composing, showing the world the way ahead in composition. His parents were very musical; talent is passed on in genetics.

Conclusions to the Learning Process
Knowing how the brain learns things, how memory patterns are laid down

for recalling, can we assist the brain by presenting the knowledge in a way most readily acceptable to the brain for storage? Yes we can. There are courses run for top executives to improve their learning technique and make better use of their brains.

How does a baby's brain develop? Cushioned in its womb, it has an evolutionary shape and inherited genetic make-up, characteristics from its mother and father, and as yet underdeveloped.

Its only likely reactions are to muscular development, instinctive reactions at birth to breast feeding. The developing baby can probably hear sounds in the womb. Music may be soothing. Mother's digestion or indigestion must sound like Niagara Falls! I wonder if at this early state the brain is trying to analyse incoming sounds? This would suggest it may first be learning through its mother's voice. In nature we see different ways babies keep in contact with their mother. After birth, vision must develop quickly. By the age of 1 year, a baby will recognise an individual face. Brain connections are putting images into memory. Some images are more important than others and levels of importance are laid down for recalling. An easy way to explain this would be the build up of levels of micro-currents. There is an enormous amount of information to be laid down and in the developing brain there is room for change. Later in life this is less possible. The influences at an early age on the brain will play a vital role in the growing individual. In the early stages, the brain's response to learning starts with:

1   Hearing.
2   Seeing – to recognise.
3   Touching – contact with surroundings – spacing.
4   Smell and taste.
5   Trying to imitate – e.g. early speech.

I demonstrated in hypnosis, regression, when I told an adult she was only five and now going to write her name. The importance of this became apparent when she displayed all the surroundings and posture recorded in this past event. The baby learning to speak is experiencing her surroundings in equal proportion to everything else. We are looking in excitement at the first utterances of Ma! Ma! The baby's brain is recording mummy with a big smile outstretched arms etc. The brain is recording all associated events.

Letters have to be learnt, connected to make words, words to make sentences, pictures to give visual memory, and so it goes on and on the art of learning, until we die. In old age many brain cells will have died. We will have less ability to process the incoming information and become more selective in our ability to analyse, but mentally still cope with our everyday surroundings.

What can we say so far about our brains?

1 It has a big memory area whose recalling gives rise to many automatic responses as well as requests.
2 Mapping out areas of the brain show areas associated with special functions.
3 Early development in childhood shows the importance of the learning process.
4 Damage to the brain gives doctors and surgeons an insight into the workings of different areas.
5 Brain scans show damaged areas in the brain.
6 Brain scans show the stages of development of a child's brain.
7 How human intelligence is different to animals.
8 What hypnosis can tell us.
9 How animals have developed their senses much more than ours.
10 The ability of the brain to perform incredible feats not possible by computer.
11 The importance of genes and biochemistry in brain control.
12 How as intelligent beings we should be able to improve our health and live longer lives and improve our sociability.
13 Learn to live within some of the limitations imposed by our brain, and look toward machines to do some thinking for us.

(h) What Can We Learn From Animal and Insect Behaviour

Do animals think? What level of consciousness do they enjoy? Can they help us to understand our own brain functions?

A lot of the senses are better developed in animals than humans. Perhaps they need better senses for their survival in their environment. These senses were in animals 200 million years ago. The crocodile has them and can live to be 100 years of age. Early man 250,000 years ago only had an expectancy of life of 18. How much could he be expected to learn in such a short time?

The output from all our senses makes consciousness. Animals are all conscious of their surroundings. Their instinctive functions are searching for food, guarding their territory and reproduction. Their ability to communicate and to think through a problem and act on it is limited. This doesn't prelude them from having emotions, feelings of affection for others. Swans display an incredible courtship dance and stay together for life. Owls stay together for life. The way in which their brain thinks is what we would describe as basic. They eye each other up as a source of food. Humans don't need to eat each other. We can hunt and grow our food. Our brains tell us to create a better environment where we can build better homes, shelter from the elements. With plenty of patience you can train animals to have emotional feelings towards humans. This is true of birds.

As I write, I see two butterflies in the garden, so small, so beautiful. Instinctively they know how to fly, to feed, and to find a mate. What a battering they are getting in the strong wind, blown here and there, always with some degree of control, finding flowers, finding each other.

Do Animals and Birds Have Emotions?

I was talking to the keeper of an Owl Sanctuary. 'Do you know birds have small brains? They are stupid. The female owl grows much bigger than the male owl. When a male owl comes to offer a female courtship, resting at a reasonable distance, bird language takes place. If the female is not receptive, the male must fly off quickly otherwise the female considers the small owl a good meal. If accepted he flies in they stay together for life, probably because they are creatures of habit and lazy, e.g. happy to sit together and do nothing, or the chances that the male might not find another mate who doesn't eat him.

The keeper of the Owl Sanctuary did not know my friend and I had reared an Indian Eagle Owl from an egg. The owl grew up believing us to be mum and dad. The owl was very beautiful but had fierce talons, and an unpredictable reaction to people she didn't like. To us, she showed emotions, love displayed by wrapping her enormous wings around our face and cuddling us. When we went away for three days she displayed classical signs of rejection. She displayed an understanding of some things we said probably as a result of familiarity with her surroundings. She displayed frustration when she tried to communicate with us, as if we did not understand her.

Crocodiles have developed and adapted to their surrounds so well, and that they still exist is amazing. They have good eyesight, can smell rotting food 4 miles away, can eat and store fat, control their heart rate, hibernate and do not need to eat for up to 2 years. They care for their young for a long time to ensure survival and live 100 years. In a cold climate they produce more females to ensure the species keep going.

All creatures on earth react to stimulation of one type or another. Sharks display mostly instinctive reactions. Their eyes are the product of development from 250 million years ago. They have a very acute sense of smell. They receive low frequency sound signals that can detect the struggling noise of a wounded fish. Where their eyes are set well back they move their head from side to side to get a good view of you. Vision directly ahead is obscured by its snout. Electromagnetic sensors on its snout help to guide it to its prey. The whale's brain size might be a reflection on its own body size. If the brain is to receive sensory signals from its skin, the surface area of the skin is enormous. I think the whale is an example where nature has allowed it to grow too big and become less efficient.

Dolphins and whales have density of neuron connections in the hearing cortex. The whale's brain is less dense than the dolphin's brain. Sound travels faster and a greater distance in water than air. The song of the whale can be heard by other whales at great distances.

Dolphins are very well adapted to their environment. What lessons can we learn from them? Dolphins don't have hands and make tools. Their advancement lives in their ability to communicate. Their speech band in communication uses ten times the range of man's hearing and the nerve connections from their ear to their brain cortex is three times as great. We have more connections in the visual cortex. The combination of their dense neuronal connections gives their brains the overall computing power. An extraordinary adaptation to their environment has occurred. Because fish bury themselves in the sand they have developed echo location. It is activated by clicks of sound energy, much like a ship's sonar that can send an impulse to the sea bottom and measure the time of the returning signal to tell us the depth of the water. We call this echo-sounding. By this process they can build up a picture of a fish under the sand that cannot be visually seen. They can sound see. The sonic sensors in their cortex are connected with the visual cortex. You can test this with an experiment. Show a dolphin an object. Now put the object in one of several boxes and ask the dolphin to find it. Every time the dolphin goes to the right box. The dolphin can see into another dolphin's body. They can see into our body and recognise us as similarly intelligent beings. The extra connections add to the computing power of the brain. The developing brain computes quicker than ours. It may be able to see another dolphin's brain at work. This suggests some telepathic communication has developed. You can test this possibility by an experiment. Tell two dolphins to invent a water acrobatic display of jumps, rolls and turns. They will produce a new sequence, one dolphin taking the lead, the other dolphin following every movement about one second behind. As a sailor I have spent hours talking to them. Talk slowly to a dolphin and he will try to analyse what you are saying, and talk back to you. Their brains have developed this sociability, the desire to communicate in a friendly way.

Genetics

When we decide to have children we don't go to the doctor and say we want to get married and have children, but I am 6 ft and have a large head, and my wife is only 4 ft and has a small head. And what will happen if our child inherits a big brain from dad and only a small skull from mum to put it in?

Nature is very accommodating but not always so. Our brains develop in a soft calcium skull which allows the brain to expand. When the brain has grown, glands secrete to stimulate bone formation in the plates of the skull. This is called ossification. What happens if ossification started too early? The brain couldn't expand any further with disastrous consequences. Does this developing brain tell the gland to start ossification, or does the gland work on a time clock? The gland works on a time clock.

What happens if we inherit small jaws from one parent and big teeth from another parent? The big teeth won't fit into small jaws so off to the dentist we go, and while we are at the dentist there may be another job to do, to correct nature. The jaws of modern man are getting shorter. With less chewing action needed on the occlusal surfaces of three molar teeth, the third molar, often called the wisdom tooth, has not room to come through. 'Hey Mr Brain, it is time you gave up making a third molar tooth, it's not needed today and I hate all these visits to the dentist'. The brain doesn't work like that.

Perhaps our brain is not so perfect. We need to modify it and make it better. We inherit characteristics from our parents. It follows we may inherit intelligent genes. Some people are more intelligent than others. That adds up! What happens if we now do genetic modifications to our intelligence, and we make the brain work better? What a dreadful scenario, that we can alter man for better or for the worse? Let's leave this line of thinking.

(i) Conclusions

I believe the following to be true. You may disagree with me.

1   There is not enough evidence to show why the brain developed so large

with the prefrontal cortex before the written word and language came into existence. It is almost as if 500,000 years ago there was a superior race, or by intervention the brain was altered.

2   I believe a contributing factor to the development of our brain has come about through the use of our hands.

3   Human and dolphin's brains are the most advanced brains on our planet. We are the only species that plays into adulthood. The added ability of dolphins to communicate by visual hearing strengthens the bonds between them.

4   Our altered environment of watching television and playing with computers is not producing the family bondship between us which we once had. The lack of communication between ourselves may lead us to rely more on increased sexual bonding as an alternative.

5   We have been given a big brain. It is up to us to use it and increase the neuronal connections. That way we will be able to find out the truth about our existence.

6   We should program our brains to receive infra-sound. We should program our brains to develop sonar, to develop sound vision. When this is well connected to our visual cortex, we will be able to do telepathic communications. In the meantime I will leave you with a different thought.

In 200 years the industrial evolution has produced a pace of life, a pattern of stress never dreamt of before. We now even catalogue stress diseases. Sailing my yacht across the Atlantic from Bermuda to England for 30 days, bathed in sunshine from the Azores high pressure in summertime, with calm seas, gentle breezes, all the stars at night, no pressure for rushing here and there, no noise from telephones, door bells, motor cars, my mind became so peaceful. I cannot describe in words the total feeling of completeness and serenity. Should we change our pace of life?

References for further reading.
BRAIN STORY *Susan Green field,*
Published by BBC Worldwide Limited, Woodlands, 80 Wood Lane, London W12 0TT

Chapter 6

Unexplained phenomenon –
ghosts, poltergeists, spirits
Unidentified Flying Objects

In science we can conduct experiments and within the parameters of knowledge, based upon action and reaction we can study the results, reproduce the experiment and get the same results, e.g. the concept that 'it is proved'. In this chapter we are examining experiences that are not always measurable with scientific instruments, but do exist and repeat themselves in various forms producing certain reactions. The existence of these manifestations and forces are registered in our consciousness as:

1 Visual (to see a ghost).
2 Auditory (hearing doors slam with no one there).
3 Force (the movement of objects as in poltergeist activity).
4 Spirits – good and evil.
5 Extra sensory perception.

The Roman Catholic Church and the Church of England recognise good and evil spirits. There is a bishop appointed by the Church of England to 'oversee' the procedures of performing exorcism to drive out evil spirits that possess people.

In the Caribbean island of Haiti there are many followers of the Voodoo Cult. The Voodoo Cult recognises good and evil spirits and cultivates the art of communicating with them and dealing with them. Rada spirits are easier to understand. Congo spirits are complex, and some are 'hot' spirits.

Shando is similar to Voodoo and is practised in the Caribbean Island of Grenada and Trinidad. In Voodoo zombies are 'raised from the dead'. Witch doctors use a voodoo poison, too much kills the victim, the correct amount causes complete paralysis. The victim can see and hear everything going on but cannot do anything. The white ground powder given to these unfortunate people contains powdered puffer fish, which contains a chemical called TTX, a powerful neurotoxin. The person is buried, as if dead, and dug up and brought back to life a short time later. These rituals of music, dancing, drug taking, exhausting the body, are designed to make your body and mind more receptive so that spirits can enter them. It is a state of self hypnosis culminating in the killing of a sheep or goat and drinking its blood.

Shando, voodoo cults are African in origin. Practising these rituals keeps the connection open to their ancestors, Africano revivalism, the umbilical line. The Incas stressed the importance of keeping the line connected to the gods in the sky, and the Inca lines on the ground in Peru are symbols of this connection.

Ghosts and Spirits can be identified as people from the past. It is suggested their transition into the next life has been suspended in a time event horizon due to traumatic circumstances, unfulfilled ambition and stress.

When an unexplainable phenomena occurs we can only observe them on as many occasions as possible, to see if there is any common ground for understanding them. One day when we have more knowledge we may be able to measure them in scientific terms. Some people can sense the presence of ghosts and spirits and even see them and communicate with them. Dogs and some animals appear to see them before we can. I think we could train our brains to be more receptive to them.

When several people see a ghost it is important to write down a detailed description given by everyone at the group sighting. Their accounts are the result of their brain interpretations of the event, which can be influenced by their surroundings at the time of the sightings, and also their previous experience of sightings. Photographic and tape recordings of the events are invaluable. The famous case of poltergeist activity in a house in Enfield in 1977 is fully documented on tape recordings and camera. In reporting a traffic accident it is amazing how many different accounts may be given. It is almost as if the brain fills in a few extra details to complete the picture.

Can philosophical thinking help us to understand these bizarre events? The philosophical approach must be considered. Philosophy can be likened to the gradual build up of scientific knowledge over many centuries. Thinking progresses forward. The German philosophers Kant and Hegel took into account the history of philosophy in its build up of knowledge.

Kant lived from 1724–1804 and Hegel 1770–1831. Hegel believed in the 'absolute mind', the highest order that is achieved, produced by the body, brain and their surroundings at that time, the theory of Monism, that there is only one state. Minds and bodies are not different in their intrinsic nature only in the way materials is made. The 'absolute mind' could then appear in an after death experience as a total replica of ourselves at that time, displayed in a different medium. Hegel took an interest in science, but rejected the current atomic theory and some of Newton's laws in favour of a theological interpretation that god made everything. I wonder how his views would have changed today with present scientific understanding? It is easy to build on wrong assumptions. How accurate are the conclusions that are reached? When an extraordinary experience happens to you it is more convincing. I have had two experiences I will recall. The first one illustrates unfulfilled ambition. The second one demonstrates the presence of life after death.

I was driving to London with my box of slides to give a lecture to the cruising association, Ivory House, St Catherine's dock. I had not been there before, and since I do not know London, driving was a nightmare scenario. In London South of the River Thames I felt an urgency telling me to take back streets and avoid the traffic, and within no time, I found myself across the Thames driving East down the embankment. When I stopped to ask directions 'where is St Catherine's dock' I was told opposite you, over there. I could see it. What a relief to have found it so easily. I was about 15 minutes into my illustrated lecture on transatlantic sailing when an elderly man in the audience stopped the lecture and said 'who is your pinta, the pilot talking through you? It often happens here'. I explained I sold my last yacht Mouette to Brian Craigy Lucas, a keen member of the cruising association, who was planning to give a lecture to the cruising association about his transatlantic adventure, but never came back, assumed 'lost at sea'. I completed the lecture. Afterwards I wished I had asked the elderly man how he knew I was possessed by Brian's spirit. Instead a stiff glass of rum and off to stay with friends in London for the night before driving back to Hampshire next day. At 1.30 in the morning I woke to find Brian standing alongside my bed. It was startling. I said to Brian, you wanted to give the lecture and I've done it in your place. I hope I did it well enough. The event is complete; you must accept this and not be troubled by your unfulfilled ambition. You must move on. It was some time before he went and I remember repeating some words again.

The second extraordinary experience happened in Jersey in the Channel

Islands. Mary worked in the bank in Southampton and retired at the age of 55 to look after her sick boyfriend and father in Lymington, Hampshire. They both died in the following year. The call for help came when Mary was dropping her gin bottle. I think her doctor wished she didn't pick up the gin bottle. Examination revealed the start of total muscular wasting. A brain scan revealed degeneration of the cerebellum a basal part of the brain. The condition progressed so fast Mary died in months. I was in charge of her ashes. Several month passed before I could fulfil her wishes and scatter her ashes in the sea in St Brelades Bay, Jersey, where Mary spent many happy holidays in L'Horizon Hotel. It was a still, cold windless November day, where the overcast sky of clouds prevented any heat reaching the ground. Others were present at the scattering of her ashes in the sea, off a small jetty in St Brelades Bay. I said the words she wanted me to say. 'This is Mary'. As the ashes nearly hit the water they spiralled up and passing the end of the jetty headed for L'Horizon Hotel at great speed. Mary's friend Susan from Guernsey said 'Mary is making for L'Horizon!' It was a real event, not imagined; nor could anyone have predicted that it was to happen. It was Mary demonstrating her presence in the next life and fulfilling her wishes to be in L'Horizon Hotel.

I went to the funeral of Dr Vernon Harrison in November 2001. Vernon was a founder member of the Liszt Society, a music society of which I am a member. Vernon was a scientist, an authority on hand writing, a music lover, and head of the London Photographic Society, and amongst many other things a member of the Ghost Society. He once told me, people phone me and say 'Vernon, I have a very active ghost; would you like to come and see it?' Vernon has photographed hundreds of ghosts. He was convinced there is life after death.

Ghosts can appear in modern houses, but typically we associate them in older buildings which can give us a scary and spooky feeling. If you are told you are going to spend the night in a haunted house, does this not make you feel apprehensive? Can we artificially create the feeling of presence? Yes, we can. The generation of sound waves too low to hear can affect the eyes and cause hallucination, that gives the feeling of a presence. Also magnetic fields can cause hallucination. As an experiment ask a person to spend a night in a known haunted room at full moon. Introduce sound waves, a draft of cold air, and the brain will interpret a presence. Does the next life operate on a low frequency of sound waves? Owls have eight times better vision than humans. Can they see ghosts before we can? Can dogs sense sound waves too low for the human ear?

Poltergeist

The word means holy spirit. Poltergeist activity has been recorded in history for a thousand years BC. Today it is as active as ever. It is thought to be associated with spirits from the dead, where terrible tragedies from the past are revived and acted out. Stress is associated with it.

Psychokinesis (PK) is the ability of a person to produce physical forces by mind power. Sometimes this can be subconscious and generate spontaneously and then it is called Recurrent Spontaneous Psychokinesis (RSPK). This could be the driving force behind poltergeist activity. There have been many experiments set up to study PK. It does exist. The group Sorrat, the Society for Research on Rapport and Telekinesis headed by Professor John Neihardt produced a large range of activity including levitation. Heavy objects move as easily as light objects. Some people have a natural ability to demonstrate PK. This was demonstrated on British television by Uri Geller, bending spoons, stopping clocks etc; also by Nicholas Williams, Nina Kulagina and Felicia Parkin can move objects by PK. This is all documented and on film. The focus is on the person who is connected with the poltergeist activity, it cannot be ruled out other forces are not at work. Certainly it is a strange phenomenon and an energy so big it has the ability to move objects far too heavy to lift.

Over the centuries poltergeist activity has attracted enormous media coverage. Poltergeist communicate and certain patterns follow, e.g. The appearance of stones being thrown, landing with force, but not causing any physical distress. Also there are tapping, loud noises, levitation, furniture moving, objects smashing, lights coming on and off, doors opening and closing, bad smells, and sometimes much damage inside a home 'all hell let loose'.

An Example of a Case

In a previous life a daughter lost her father in a tragedy and was deeply upset. The present family living in the same house had a similar experience. The daughter was deeply distressed when her father left in a divorce situation. This triggered the spirit of the deceased to work through the daughter living in the same house. The spirit spoke in their voice through the daughter's voice box.

Ways of dealing with the phenomena vary from case to case. It is necessary for a medium to communicate with the spirits, understand their problem and help to drive the spirits into the beyond where they have been held back by the trauma of past events.

There are medium and exorcists who claim to be able to see these spirits, talk to them, and understand their problem. Maybe if we knew how to use our brain better we could all develop the ability to communicate.

The experience that Verona Seiter had is an example of psychokinesis. I was learning to do energy work with an energy teacher. There are blocks of energy in a person and when you look at a person you learn to look at their energy pattern. You look for where the energy is blocked. You massage that particular part of the person's body.

The first unusual thing that happened, my teacher was working on another person and I was standing watching her. She asked him if he felt any fear. He said 'no'. Then she asked him if he felt any anger. He denied that also. At that moment I started violently shaking from head to foot. My teacher looked at me and said, 'let it go, it is not yours'. She told me that you have to let the other person's bad energy channel out through you, and let it out, do not let it hang onto you.

Then I started waking up in the night, sitting up and speaking in tongues; some tongues were old, some were young, speaking many languages through me. I could not stop it happening. It was no dream. This happened many times over a two week period. It scared me. During that same time period I was massaging my daughter's neck. I did not think it counted as body work so I did not wash my hands up above my elbows in cold water. What happened next was scary. In the middle of the night I got blisters all over my body. These blisters did not hurt, then I went and washed my hands up to my elbow in cold water and said a prayer 'what have I done wrong?' I went back to sleep. When I woke up in the morning the blisters had gone, leaving some red marks.

I went back to see my energy teacher and told her what had happened, and how scared I was. Speaking in tongues was still going on. She sent me to a Medium. I had a session with her. She said some forces were trying to take over my body and I should go to a healer. In the Medium's house I was causing the doors and windows to rattle. The Medium sent me back to my teacher, who referred me to her teacher. I went to him still being sceptical about everything. He told me to close my eyes. You could feel his energy in his hands. He spoke softly and I felt a white light, very warming, go through me. This apparently took some time, but I was not conscious of it taking much time. Then he said 'sit up, it is finished! It worked; I never had any more trouble, never spoke with any more tongues. I had been training for several months with about nine other people. I never went back, and stopped doing energy work.

Unidentified Flying Objects = UFO's

Also known as flying saucers, have been recorded in history for thousands of years. Yes, we are being watched. The difficulty is we do not have their technology. How do they fly? How do they communicate? They seem less interested in communication, more interested in observing us, possibly experimenting on us, and collecting soil samples. If they live on an artificial platform perhaps they do not have soil. I wonder how many different types of aliens have visited us, and whether they are all friendly. They appear to arrive in very large spaceships and flying saucers come out of the mother ship to land on earth.

Have we seen aliens, have they landed, have we talked to them? Do we have crashed parts of their craft? What is it made from? What is the alien blood chemistry? How do they differ from us? The simple answer is yes to most of these questions. There has been a cover up. Certain persons have been told they will lose their jobs if they divulge knowledge gained at this very top level of secrecy. In the arms race between superpowers military advantages could have been gained by the extra knowledge.

Sightings

They have been recorded by artists in paintings which can be seen today. The miracle of the snow painted in 1429 is in the Capodimonte museum in Naples. It shows lots of UFO's. What else can they be? The Madonna and St John (1450 period) at the Palazzo Vectio in Florence has a UFO being watched by a man and his dog. In 1561 numerous objects were described in the sky over Nuremberg, some low down and travelling at great speeds. In 1896 several objects were seen over Sacramento, California, and numerous reports came in of sightings in America and Canada.

On 25 February 1942 a UFO appeared over Los Angeles. Seven searchlight beams converged on the object and anti aircraft shells burst on the object with no effect. This was photographed by a Los Angeles Time's reporter. Numerous sightings occurred during World War II. Radar trackings were reported and aircraft sent up to intercept. Pilots described what they saw. Giving chase they were never able to intercept the UFOs which would accelerate away at enormous speed. Their reports include seeing high speed craft not looking like conventional craft.

The George Adamski Report

In California in 1946 George Adamski saw a leviathan in his telescope. Hundreds of people saw it and the San Diego radio carried a story about it.

123

He photographed a leviathan in March 1951 and again in 1952. These are cigar shaped with blunt ends and what appears to be a row of cabin lights.

ADAMSKI SAUCER

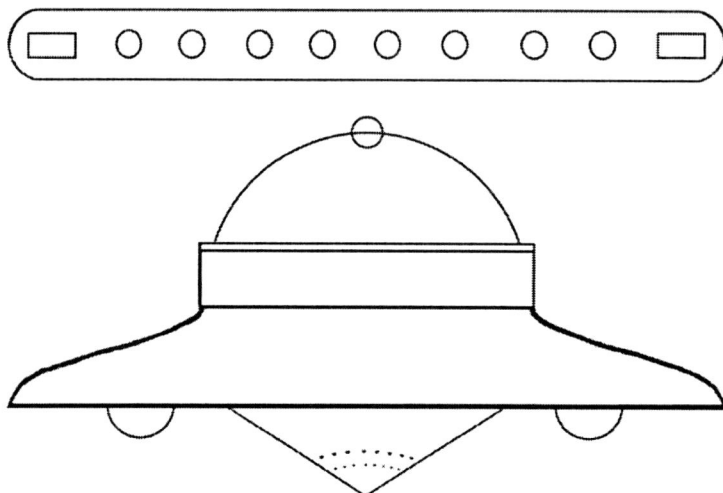

He wrote books 'Flying saucers have landed (Leslie and Adamski 1953) a large presence of UFO's in 1947 may have been due to the explosion of atomic nuclear bombs in Japan. They have come to see our technology. Perhaps we are reaching the level of knowledge that interests them. On 8 July 1947 a flying saucer crash landed on a farm near Roswell. It was witnessed, pieces of wreckage examined by ordinary people, before the US Government ordered the Roswell Army Air Force base in New Mexico to seal off the area They issued a press release saying the wreckage was a weather balloon. This was the first big cover up. The real wreckage, to- gether with dead aliens, was taken to Wright Patterson Air force Base in Daytona, Ohio, for analysis. The aliens were described as small, large heads, big eyes, no body hair. Two more craft crash landed in New Mexico in 1948 with aliens aboard. One craft was not badly damaged and one alien was still alive. Our chief Marshal, Lord Downing, commander in chief of RAF fighter command during the Battle of Britain said 'more than 10,000 sightings have been reported, the majority of which cannot be accounted for by any scientific explanation. I am convinced these objects do exist and that they are not manufactured by any nation on earth'.

Admiral of the Fleet, The Lord Hill-Norton said 'the evidence is now so consistent and so overwhelming that no reasonably intelligent person can deny that something unexplained is going on in our atmosphere'.

Wilbert Smith, in a 1950 top secret Canadian government memorandum said 'the matter is the most highly classified subject in the United States government, rating higher even than the H bomb. Flying saucers exist!'

In 1967 Professor Felix Zigel of the Moscow Aviation Institute, Moscow Central Television, said 'unidentified flying objects are a very serious subject which we must study fully. This is a serious challenge to science and we need the help of all Soviet citizens.'

Colonel Boris Sokolov, Soviet Air Force, confirms a state funded UFO search project was initiated in 1977.

Many of the government reports on UFO sightings have stated, 'in not so many words', 'the incidences have been investigated and found to be of no military importance', e.g., no defence importance. The statement could also be replaced by alternative words. 'The incidence has been investigated and found to be of scientific importance'. How can a possible alien event come under any government official secret act. Reports of UFO's buzzing airports brings in the Civic Aviation enquiries, which do not dismiss the possibility of an alien explanation.

Mrs Thatcher was asked for her opinion. She said 'it is not in our interest the public should know'. When hundreds of people observe the same thing it is difficult to deny its existence. When military personnel observe the UFO it is accepted their account is reliable, e.g. the Rendelsham Forest case 26 December 1980, when a UFO landed in a woodland east of Ipswich near two NATO bases. The landing sight next day showed 'landing gear' prints in the ground and high radiation levels. There was a further landing in the same woods two days later when aliens were seen. The incident was officially reported by a letter dated 13 JANUARY 1981 from the Ministry of the Air Force, but this letter was not released until 1984. The Ministry of Defence has now released the Rendelsham files to public view. Deputy Base Commander Lieutenant Colonel Charles Halt of the US Air Force Nuclear Base at RAF Woodbridge reported the incident. He told how three military policemen saw lights in the trees outside the back gate of the airfield and set off, fearing a crash. They reported seeing a strange glowing object in the forest described as being metallic in appearance and triangular in shape, approximately two to three metres across the base and two metres high. It illuminated the entire forest with a white light. The object itself had a pulsating red light on top and a bank of blue lights underneath. The object was hovering or on legs. As the policemen approached it manoeuvred through the trees and disappeared. At this time the animals on a nearby farm went into a frenzy. The next day Lt Col Halt joined a party which found three depressions on the forest floor where the object had been sighted. Radiation readings of ten times the normal level were found around the site. The strange lights returned after nightfall. This was recorded on a Dictaphone. The Officer was heard saying 'there's no doubt about it, there's a strange flashing red light ahead, pieces of it are shooting off, this is weird, it's like the pupil of an eye looking at you, winking'. After a period of time the observation continued. 'We got two strange objects half moon shape darting about with coloured lights on them .... Yeah, they're both heading north. He's coming in toward us now now we're observing what appears to be a beam coming down to the ground. On the dictaphone there are excited shouts from members of the four man patrol and another officer explains 'look at the colours shit! This is unreal gasps Lt Col Halt. Other patrolmen backed up the story. Larry Warren said these were not just lights but something far more incredible. A triangular object appeared right in front of him and he felt nauseous as the hairs on the back of his neck stood on end. He says he saw three aeronaut entities communicating telepathically with a senior officer. The aliens were three feet tall and resembled kids in snow suits. He said they floated in a bluish/gold ball of light out of their craft. Next morning Warren said they had to sign statements saying they only saw 'unusual light'.

If you see a UFO write down the details of the observation. Remember light from the visible spectrum reaching your eyes may not be continuously omitted, so the spacecraft may disappear from view.

How do UFO's work? It is suggested:-

1   Traditional rocket fuel is not used – how we send satellites into orbit.
2   The whining noise suggests electric motors are started and the production of electromagnetic forces.
3   The use of superconductors operating at low temperature.
4   The production of laser light.
5   The production of synchrotron radiation and Cerenkov radiation.

This suggests the production of very high energy levels in the electromagnetic spectrum, which are generating energy sources from particle physics. The different lights emitted from the UFO's suggest synchrotron and Cerenkov radiation. The effect of the electromagnetic alternating current produced by the UFO's on car ignition systems, and other electrical systems, causes cars to stop and lights to go out. This does not affect diesel cars.

Alien Abduction
There are stories people have been abducted into spacecraft. Aliens paralysed their victims with light guns, and experiment on them.

Telepathic Transfer
Is it possible our brain could be influenced by aliens? I believe so. In poltergeist activity spirits work through other people's brain. A few people can demonstrate telepathic transfer e.g. Uri Geller.

Telepathic Automatic Writing
If spirits can work through our brain then automatic writing can occur. This appears to have happened to Mavis Burrows in 1982 in the small Suffolk town of Hadleigh. She received instructions to write and compose some 'space art'. The results look like something from a different world. There have been many strange tales from many people on automatic writing.

Crop Circles.
They can be very large and complex. How would humans make such complicated designs overnight? There are no. radiation effects that can be found. In crop circles we need a photograph of an alien making a crop circle. There is little to support alien intervention but keep an open mind.
   Please read the following interesting books for more information:-

1   UFO Quest, in search of the mystery machines, by Alan Watts. First published in the UK 1994 by Blandford. A Cassell imprint, Cassell plc, Wellington House, 125 Strand, London WC2R UBB
2   Open skies – Closed Minds by Nick Pope. First published 1996 Simon and Schuster UK Ltd. Africa House, 64–78 Kingsway, London WC2 B6AH.
3   Beyond Top Secret by Timothy Good. First published 1996 by Sidgwick and Jackson, then in 1997 by Pan Books, Macmillan Publishers Ltd, 25 Eccleston Place, London. SW1W 9NF
4   Alien Contact – The First Fifty Years by Johnny Randler Published by Collins and Brown 1997, London House, Great Eastern Wharf, Parkgate Road, London SW11 4NQ.

5 Poltergeist Phenomena by John and Anna Spencer – Headline Book Publishing, A Division of Hodder Headline PLC, 338 Euston Road, London NWI 3BH

6 The Encyclopaedia of Ghosts by Daniel Cohen – Brockhampton Press, a Division of Hodder Headline PLC Group.

Chapter 7

The progress of sciences throughout time
Conclusions reached so far

Most inventions have been attributed to the Sumerian, Egyptian and Greek Civilisations. Houses were being built and community settlements by 7,000 BC in Mesopotamia. The earth was being populated by Homo Sapiens 200,000 years ago, and by 40,000 years ago intelligent man was roaming the earth from China, Europe and South America. Hunting skills meant the need for primitive tools and spears. These skills developed independently in different parts of the globe, early findings are centred around Mesopotamia, lush pastures on the banks of rivers like the Tigris and Euphrates. The first cities were built here.

Approximate times, including a few historical dates.

40,000 BC	Cave paintings found depicting animals
20,000 BC	Cave paintings including humans and hunting scenes. Hieroglyphics-pictures to represent words and ideas.
12,000 BC	Harpoon invented. Tombs built.
6,000 BC at Jericho	Bricks were baked for the construction of houses, temples
5,000 BC	Bronze and copper invented, and other metals, to be used in construction. Saws were probably made of bronze. The wheel was invented, made of wood and a metal axle. Crops were cultivated. The Chinese were sowing rice. Temples and Ziggurats built, with astronomical alignment.
4,000 BC	Pyramids, stone circles, temples, built in Peru, Egypt, parts of Europe and the far east. Animals were domesticated. Harnesses were invented to put on animals to drag carts and ploughs. First sailing boats used on the Nile and other rivers as a means to cross a river and for transport. Sun Dial

	invented to keep the time. First cities built with water courses and drainage systems. Large farming communities settle around the river Ganges in India.
3,500 BC	Sumerian Civilisation invented writing and parchment. (paper was invented in China in AD 105).
3,000 BC	The day was divided into 24 hours
2,800 BC	Building Stonehenge, England, and Pyramids commenced in Egypt. Egyptians using hieroglyphic scripts.
2,500 BC	Soap was invented by Sumerians and use of herbal remedies in medicine.
2,500 BC	Climate change and Sahara Desert drying up.
2,000 BC	Minoan palaces built in Crete displaying advanced building features for the period and ceramics.
1,500 BC	Chinese make silk fabrics
1,300 BC	Greeks invented mechanical tools and iron products
1,290–	Rameses II, Pharaoh of Egypt searches for the book of Troth and
1,213 BC	the Benben stone.
907 BC	Solomon becomes King of Israel and starts the building of Jerusalem.
776 BC	Olympic games started in Greece
753 BC	Building of Rome started.
800–700 BC	Homer writes the Odyssey and the Iliad..
605–562 BC	The Hanging Garden of Babylon – one of the seven wonders of the world
604 BC	Nebuchadnezzar King of Babylon.
528 BC	Buddhism.
520 BC	Confucism and Taoism founded.
520 BC	(the golden age of moral thinking).
450 BC	Development of art and architecture in Greece.
400 BC	Hippocrates opens medical school in Greece.
334 BC	Alexandra the Great conquers Asia
290 BC	The great library of Alexandria, Egypt.
285 BC	The Pharos lighthouse at Alexandria – Development of great trading at sea.

469–399 BC	Socrates	Theories on many subjects including medicine
427–347 BC	Plato	
384–322 BC	Aristotle	
221 BC	Building of Great Wall of China.	

130

338 BC	Early Roman coins. The Roman world dates from 750 BC–500 AD.
215 BC	Roman armies in Spain.
147 BC	Roman armies in Greece.
100 BC	Glass invented
46 BC	Caesar becomes Roman dictator.
4 BC–30 AD	Life of Jesus Christ.
AD 50	Nazca lines and figures appear in Peru.
AD 79	Volcano Vesuvius destroys Pompeii in Italy.
AD 100	Romans invent concrete.
AD 330	Building of Constantinople.
AD 105	Paper invented in China.
AD 190	Galen writes first medical book and establishes theories on how the body worked, e.g. 'The circulation of the blood is like the ebb and flow of the tide.' It was not until the invention of the microscope that Harvey in 1645 proved the circulation went all the way round the body.
AD 200	Height of the Roman Empire.
AD 220	End of the Han Dynasty in China.
AD 300	Christianity became widespread.
AD 436	Roman troops leave Britain.
AD 460	Vandals destroy Rome – Decline of the Roman empire.
AD 570–632	Life of Muhammad – start of Islam.
AD 559	Chess invented in India.
AD 865–74	Vikings conquer North East England.
AD 927	Athelstan becomes first King of England.
AD 500–1000	Are referred to as the Dark Ages – wars. Paper was expensive and only for the rich. Very few people could read and write. The big advances in knowledge started in 1400 with the first universities
AD 1405	Germany used metal screws.
AD 1436	Gutenberg invents movable type for printing.
AD 1456	First bible was printed in French.
AD 1492	Great paintings by Leonardo Da Vinci – the Last Supper – The Mona Lisa. The period 1400–1600 are referred to as the Renaissance period – great architecture, art, culture, microscope and telescope and first attempts to demonstrate planetary motions also great sea voyages of discovery mapping the world.

AD 1512	Copernicus – states the earth and planets orbit the sun.
AD 1470	Muvineb Astrolabe invented forerunners of the sextant (1752).
AD 1492	Columbus discovers the new world – great sea voyages.
AD 1497	John Cabot searches for North West Passage to the Pacific Ocean
AD 1498	Vasco Da Gama reaches India.
AD 1499	Vespucci sails to America.
AD 1500	Wallpaper was invented.
AD 1511	Watch invented in Germany.
AD 1519–22	Magellan completes circumnavigation.
AD 1545	First microscope built and later refined by Galileo in 1590.
AD 1564–1642	Galileo studied physics and mathematics producing telescopes, microscopes, thermometer. He described planetary motion and the Roman Catholic Church put him in prison and house confinement for his last 10 years for having these evil thoughts! A Heretic?

Continuing wars around the world led to scientific inventions. The Chinese invented gunpowder in about 900 AD and the cannon (a large gun) about the same time.

By 1350 it was the policy of Kings of England to grant favours to merchant men with fast sailing frigates who could provide ships for defending ports, some of these rich ship owners were pirates, and now they were being given estates to manage, and castles to build for defence of sea ports. At the entrance to the Solent between the Isle of Wight and Lymington there is a spit of land called Hurst. In 1541 the foundations were laid for the building of a Kings palace, which subsequently became Hurst Fort. It was strengthened and extended to house 38 cannons to keep the French out. These were breach loading cannons. With the advancement of science chemical cordite was invented which enabled these cannons to be fired quicker than the French cannons.

| 1558–1603 | Elizabeth I becomes Queen of England. Mercator makes a new world map based on the Mercator's projection. Trade was expanded by sea routes as well as overland. |
| 1588 | Spanish launch an armada of ships against England. |

132

1623	The first mechanical calculator was made in Germany.
1637	First umbrellas made.
1661	First bank notes were made by Bank of Sweden.
1665	Great plague of London spreads across the country carried by rats.
1666	Great fire of London.
1642–1727	Isaac Newton – a great scientist who explained the laws of gravity and split light into many colours.
1709	Piano was invented in Italy.
1712	Thomas Newton invented the steam engine 1663–1729).
1741	Measurement of heat – centigrade scale used in thermometers.
1756–1791	Mozart played the piano and composed.
1728–1779	Captain James Cook made 3 voyages to Australia and the Pacific – English sailors nicknamed 'Limeys' for drinking lime juice to stop scurvy.
1769	James Watt patented the rotary steam engine.
1769	Large commercial machines invented for weaving.
1800	Volta invented the battery
1821	Electric motor and electric generator invented by Michael Faraday.
1864	The first steam locomotive was invented by Richard Trevithic. Robert Stephenson's steam locomotive could reach 40 mph.
1830	The first steam passenger/freight railway was opened between Liverpool and Manchester.
1840	First postage stamps in Britain (Penny Black).
1856	First internal combustion engine developed in Italy.
1859	Charles Darwin published 'The Origin of Species'.
1862	Alexandra Parkes invented plastic.
1865	Advances in antiseptic and general anaesthetics.
1866	Dynamite invented.
1876	Alexander Graham Bell invented the telephone.
1879	Light bulbs invented.
1884	First underground railway operated in London.
1885	Motor car invented by Karl Benz.
1885	Louis Pasteur, chemist, produces vaccines.
1895	First public cinemas.
1895	Wireless telegraph invented by Marconi.

The industrial revolution:-
All these inventions led to machines in factories that could do several men's work. These led Britain into a period of great wealth, the old colonial days. Men were prepared to take that knowledge and skill abroad to develop other countries resources and labour, although not always popular. The slave trade was profitable. Merchantmen in Bristol made big fortunes taking machinery to Africa, Slaves to the West Indies to work sugar plantations and farms, and sugar, spices and fruits to come home.

1900	Sigmund Freud published books about dreams.
1903	The Wright Brothers flew the first aeroplane.
1905	Albert Einstein developed his Theory of Relativity.
1909	Chromosomes discovered which are connected with hereditary.
1911	Rutherford explains the atom and its nucleus.
1915	Cables laid across the Atlantic for the first transcontinental telephone cables.
1915	Ford made farm tractors and cars
1920	Radio broadcasting.
1922	Medical research into diabetes isolates insulin.
1926	Penicillin discovered by Alexander Fleming – first antibiotics produced.
1901	Wilhelm Rontgen received the Nobel Prize for physics for discovering x-rays.
1903	Marie and Pierre Curie received the Nobel Prize for work on Radium.
1926	John Baird produced the first TV image.
1927	The Solvay International Conference on Quantum Physics.
1928	First TV Broadcast. First coloured TV was in 1953.
1928	The principles of a jet engine were invented by Frank Whittle. Air is sucked in a gas turbine produces a lot of hot gases which are ejected, producing thrust. It was not until 1952 that the first British Commercial Jet Aircraft was operational. The latest commercial jet aircraft Concorde can fly at twice the speed of sound.
1929	Electron microscope invented by Max Knoll and Ernst Ruska, to allow scientists to see small objects.
1935	Radar invented by Sir Robert Watson – Watt.

1938	First photocopying machine invented by an American, Chester Carlson. First colour copier invented in Japan in 1972.
1939	Helicopter invented by Igor Sikorsky
1942	First nuclear reactor built in America.
1942	Wernher Von Vraun invents rocketry.
1944	America tests atomic bomb.
1945	First electronic computer.
1948	The transistor radio is invented.
1953	Structure of DNA discovered.
1953	Polio vaccine developed.
1954	America tests H bomb.
1956	John Bardeen, William Shockly and Walter Brattain receive the Nobel Prize for Physics for inventing the transistor.
1956	Video recorder made by Ampex in the USA – It is an electronic device and records TV programs on video tape.
1957	Russia first to put satellite to orbit.
1958	Silicon chip invented.
1960	Laser invented by an American Theodore Maiman. A thin beam of very powerful light is used in industry and medicine for various purposes, e.g. precision cutting.
1961	Yuri Gagarin first man in space.
1962	Lev Landan received the Nobel Prize for Physics for his research on liquid helium.
1965	Pictures of Mars sent back from Space Probe.
1969	First moon landing.
1975	Soviet and US dock in space with joint program.
1977	The first personal computers – with visual display unit called Apple 11..
1979	Sony made a personal stereo powered by batteries called a Walkman. Earphones allow only you to hear it.
1980	Smallpox eradicated.
1981	America using a reusable space shuttle.
1983	Barbara McClinktock won the Nobel Prize for her work on genetics.
1984	Aids Virus identified.
1984	For crime – technique for identifying people by their unique DNA pattern is pioneered.

1990	Computers are linked via telephone lines to create the Internet.
1990	Interactive CD Roms are made.
1999	Russian Space Station Mir is abandoned.
2002	Human DNA Code completed.

I AM GOING TO ANALYSE CONCLUSIONS I HAVE ARRIVED AT SO FAR

1 There are laws governing the universe.

2 The common denominator is the beauty and diversity of creation, and the inherent tendency for chemistry in nature to return to the neutral or stable form.

3 Everything is made in the stars, and cloud gases in the sky have been analysed as containing the chemicals, amino acids the building blocks of life.

4 Where there is water on a planet, life forms will exist, some more advanced than others, and the universe is studded with life. Planet earth is relatively new in the universe.

5 That our brains can be accessed through psychokinesis and may have been altered and modified in the past.

6 There are forces acting on a level we cannot access.

7 That everything may have been made on a computer, and later built from material elements.

8 The way ahead is to build super computers to take over from our limited brains.

9 Scientists are spending vast sums of money on space exploration and trying to contact other civilisations. I believe we will make contact with other civilisations in the next one hundred years.

Further Thoughts:

1 We do not know all the laws of the universe and astronomers are puzzled by their findings

2 There are basic building blocks that make everything. What do we do if we make something beautiful? We want to share it with others. The beauty and diversity of life provides the purpose behind creation.

3 Creation has been present for a very, very long time.

4 The fundamental ingredient for organic life as we know it is water. In the universe planets with water may have life.

5 I can find no concrete evidence why our brain enlarged with such a complex

136

convoluted cortex. In the absence of a strong stimulus for it to expand, I can only assume there has been intervention by the designers. How much easier it would have been to delay the secretion of a hormone from the anterior lobe of the pituitary gland to delay ossification of the soft calcium skull in a child, which would have allowed the brain to expand without having to develop deep folds to increase its surface area. The later designers obviously could not find the earlier blue print. In nature can we find random growth without stimulation? Take the case of a tree in the middle of a field with plenty of water and nutrients and no competition for light. Why does the tree only grow to a certain size? Has it inherited a size and form that is modified by its need to survive? If one branch wants to grow bigger does it have to send a signal to the trunk, make yourself stronger to support my weight and while you are at it make your roots in the ground longer? Perhaps the trunk sends a signal back, 'I am doing a balancing act already with branches all the way around me, so do not get out of hand and grow longer'.

6 Our senses are not that well developed. Owls have much better sight and hearing than humans. They only have bird brains. We cannot hear infra sound so we cannot access signals on that level. We can measure signals by other scientific means. In the case of dark energy which accounts for 70% of the universe we cannot access it by our brains and we cannot measure it by scientific instruments, and yet it exists. What do we know about dark energy?

1   It has negative energy, anti-gravity.
2   It is not uniformly distributed in the universe, because nothing is.
3   It is in different concentrations producing different effects.
4   It could be artificially made by other civilisation.
5   It repels matter where its force exceeds that of gravity.
6   It invades space where there is no matter.

There would be a boundary between matter and anti-matter where the forces cancel out. Can we observe it at work? Yes, it might be found near small high density galaxies where there are no peripheral stars, indicating it has enough energy to repel the most distant stars from the centre, but nothing nearer, like a boundary layer.

8 Was everything designed on a super computer? It could have been. Why are there nearly 200 subatomic particles? Can intelligent life be very small? What other life forms are there in the universe?

9 Building new equipment to study science is the traditional way ahead and, step by step, inventions open up new lines of research. Do we have to explain everything in Quantum terms?

10 We must explore space, that's why we have been given bigger brains. If we can find signs of life on planet Mars we can say we are not alone in the universe. Two rockets and probes are on their way to Mars to probe the Martian surface, and we will have news of their findings by the end of 2004.

Chapter 8

Religion

Early religious rituals were practised 8000 BC. Decorated skulls have been found in Jericho dated from these times. A long time before that I expect people worshipped their own Gods, the God of the sun, the God of the moon, The God of Thunder and Lightning, the God of the Rainbow, the God of Storms, the God of Evil and Fire. Did they believe these Gods lived in the sky? The Incas, much later, did.

The Greeks worshipped many Gods. Zeus was the King of the Gods, Poseidon was the God of the Sea. Aphrodite was the Goddess of love. These Gods were believed to live on Mount Olympus which is in Northern Greece.

Egyptian Gods were numerous. Horus was the God of Light; Maat the God of Truth, Ra the God of the Sun, Thoth the God of Knowledge. There were Norse Gods and Goddesses. Thor was the God of Thunder and war. In addition there were Mythical creatures with animal and human features. The Sphinx had a lion's body and female head. Centaur had a male torso with horse's legs.

What distinguishes religion from other beliefs from moral codes, magic, ancient custom and rituals? Is it a belief that the absolute goodness overcomes the evil in the world? Is it witness to a divine intervention, an experience of miracles? Is it a function of our brain in the balance of neurotransmitters sending messages along our brain cells?

Religion is an encyclopaedia. We need to study all the religions to find a common denominator. The poor and needy may turn to religion for help. For a solitary man there may be a need to find comfort in a supreme being, and in bad health and dying the need to turn to someone for help. To quote SCHLEIERMACHERS 'The essence of religion consists in the feeling of an absolute dependence'. In Islam children are made to learn the KORAN from an early age, a form of indoctrination, when their reasoning powers have not developed to make comparison. This is the reason why it is the fastest growing religion.

What is the proof of the existence of God? Can we in any words in any language describe an all powerful good God? Could we generalize to include in religion unexplainable scientific phenomenon or extrasensory perception, medium, ghosts? Apparitions from beyond the grave may be a natural phenomenon which occurs across an event horizon or as Aristotle says 'The development of the PSCHE which lives on into the next life'.

Are we all connected to a giant computer? I am writing a guessing game. Where is God? We don't know. Christianity says the HOLY SPIRIT is God's continuing presence in the world.

Let us examine what different religions tell us. The earliest recorded religion, when a claim is made that God sent a prophet to earth, is Judaism, in 2000 BC. God spoke to Moses, and Abraham made a covenant to the Jewish people.

The beliefs are based on the Old Testament of the Bible.

1   One God only.
2   God made man in his own image
3   They believe a Messiah will come one day to redeem mankind.

Judaism has some flexibility to the original strict orthodox Judaism. Conservative Judaism is open to many interpretations. Reform Judaism goes a stage further and says you can modify laws and traditions. The Synagogue is the house of worship and Jewish affairs. Jewish communities are scattered across Europe, America and Israel.

The earliest recorded teacher/prophet in Central America was Quetzalcoatl, of 950 AD. He came 'down from the sky' with fair skin and red hair, taught the people in Mexico. He knew how everything was made. A temple exists to him at Tenochtitlan. His fame spread far and wide, and as late as 1530 AD he was remembered by the Incas who mistook the Spanish for Quetzalcoatl's second coming. He rose from the dead, the prowess of resurrection.

In 1500 BC Hinduism was founded. Scriptures were written called the Vedas, stressing the consequences of our actions in this life are taken into the next life. There are many Gods and sects, so the religion is diverse and taught by Brahman priests. India is the main centre of this religion. There are many symbols to help your meditation and offer prayers

Brahma is the Hindu God
Vishnu is the God of Love.

Om is the sacred symbol of spiritual enlightenment.
Swastika symbolizes good luck.
Shri Yantra symbolises wholeness.

600 BC ZORUASTRIANISM
Founded by a Persian prophet, ZOROASTER; it has a holy book the
AVESTA and a set of beliefs:-
There are two forces – good and evil.
1   Good will triumph
2   The dead will be resurrected
3   There will be a paradise on earth. It was widely practised for some
    centuries, but little today. It has been superseded by HINDUISM.

600 BC TAOISM
Founded by Laozi. It was practised in the Far East in China. It is really a
philosophy not a religion. It believes in controlling spirits by:
1   Balancing forces of nature to achieve perfect harmony.
2   Meditation and simplicity.
3   Seeking immortality.

660 BC SHINTOISM
Founded in Japan. Two main traditions
1   Shinto Shrine – JINGA – National religion under 1945
2   Sectarian KYOKO – Group leaders and teachers.
Sacred texts are the ancient records of Japan.
Sacred Shrines are the sacred gates, the holy tree etc.

599 BC JAINISM
A monastic system; an offshoot to HINDUISM founded by
VARDHAMANA MAHAVIRA
– A high moral code for people to follow.

551 BC CONFUCIANISM.
Founded by China by Confucius. It is a good moral code, more a philosophy
than a religion.
It states: -
1   Good code of behaviour in conducting public life.
2   Good code of behaviour in private and social life.
3   Good character that can be respected to set a good example.

4   Respect for rulers and elders.
5   To become better through learning and self examination.
6   To become a good family man.

300 BC
The Golden age for China, the combination of TAOISM and CON-
FUCIANISM and at a time when the Chinese States became independent
from the feudal lords over which the old Chow dynastic power had
presided.
   Taoism stressed the need to balance nature with man's achievements, to
have harmony and return to natural simplicity. This leads to a long life and
better understanding.

528 BC BUDDHISM
Founded by Prince SIDDIIARTHA GAUTAMA in India (563 BC–483
BC). A Rich man went out to help the poor. Buddha = the Enlightened
One. Buddha says:-
1   Suffering is present.
2   Desire causes suffering.
3   Suffering ends with desire to overcome.
4   To achieve NIRVANA
Indian class System:
1   RAHMSANA – The privileged priest class.
2   KSHATRIYA – The ruling warrior class.
3   VAASYA – The merchant and land owning class.
4   SUDRA – Servants.
Spread of Buddhism.
The Indian Emperor ASOKA became a Buddhist and promoted the faith
throughout his empire.
AD 50 reaches CHINA
AD 550 reaches JAPAN
AD 600 reached TIBET

AD 600 ISLAM

Founded when the prophet Muhammad was given a new faith. The Koran
is the sacred bible. They pray 5 times a day while facing the holy city of
Mecca and fast during the months of Ramadan. Muslims are the followers
of Islam. By 750 Islam occupied an Arab Empire from Persia across North
Africa to Spain. In its early concept Islamic philosophy remained simple

and easily understood, and concentrated on Syncretism (blending of inharmonious elements). Muhammad was influenced by Greek philosophy. Later Islam became involved in other neighbouring countries religion, including Hellenism Christianity. Islam says there is one God and Muhammad is his Prophet. Muslim thinkers wrestled with the same problems that their Christian counterparts wrestled with, namely freewill, predestination, anthropomorphism (man's wisdom in understanding the universe and spiritual perceptions) and allegory and divine justice. In their search for the truth Neoplatonism played a part, Aristotle's reasoning and ideas of Plotinus. Islamic theology has a mixture of Greek, Neoplatonism and Christianity. Muslims do not trust the West. There is a bridge to be crossed. The indoctrination of children at a young age with learning the KORAN keeps the faith alive and spreading. The individuals right to go to war if he feels justified 'in the name of Islam' is disturbing. Muslims have made some good technical advances in the early years, capturing the art of paper making from the Chinese around *750* AD. This is how papermaking came to Europe.

AD 1 CHRISTIANITY

Founded by Jesus Christ 'The Messiah' 4 BC–30 AD.
He rose from the dead and showed himself to his disciples. The 12 disciples spread news of his resurrection. Christianity is based on the life and teachings of Jesus recorded in the 4 gospels of the new testament. It started in the Eastern Mediterranean and eventually by 400 AD was accepted in Rome by Emperor Constantine, after the Christians had been persecuted. St Peter, a disciple of Jesus, arrived in Rome in AD *50* and eventually the Roman Catholic Church was set up with the Pope, the head of the church. St Paul was active in spreading the faith across Syria, Arabia, Greece and Rome. St Paul lived from AD 3–68.

Teaching and beliefs.
Jesus is the son of God and demonstrated life after death by resurrection. God is known as father in heaven. The Holy Spirit is God's continuing presence in the world. The sacred texts are the Old Testament and the New Testament. Jesus will return at the time of the second coming. The Roman Catholic Church and higher churches use objects associated with prayer. (1) Crucifix. (2) Religious paintings. (3) Rosary beads.

In 1054 there was a great disagreement between the Christian Churches of Rome and Constantinople. The Pope excommunicated the Patriarch of Constantinople, and Christianity continued as the Eastern Orthodox Churches. There are 14 churches governed independently under its Patriarch. The churches are divided into dioceses and parishes. Parishes are led by priests and deacons.

In 1095 Pope Urban II encouraged Western leaders to recapture the holy lands of Western Asia from the Muslim Turks. This was the start of the Crusades which lasted 108 years. There were four crusades. The Muslim warriors were called 'SARACEN' and CRUSADERS went to battle with a red cross on their uniform to show they were fighting for GOD! The Crusaders did capture a lot of land back from the Saracens but not the holy land.

In 1534 The Church of England was set up as result of Henry III break with Rome. Some sections of the Church of England follow Catholic traditions, other stress Protestantism (high and low church). The Archbishop of Canterbury is the head of the Church of England. Dioceses are governed by Bishops, Parishes by Clergy. Other Protestant Churches:

1530   Lutheran Church by Martin Luther after split with Rome.
1604   Sikhism founded.
1609   Baptist Church founded in Amsterdam
1612   Baptist Church founded in London.
1648   Quakes founded.
1729   Methodist Church founded.
1830   Mormons Church founded by Joseph Smith.
1865   Salvation Army.
1866   Christian Scientist.

Conclusions

A lot of religions were started in the period 500–600 BC. I do not know why. It would appear it could be the influence of religion on India and the Far East, China and Japan. China was the most developed country in the world at this time. World disasters can cause religions to start. When Krakatoa erupted in 535 AD the polluted atmosphere blocked out the sun for 5 years. Photosynthesis was much reduced and crops failed. Some people died of starvation, but by 550 AD most people had died from the plague. Was it a coincidence the Islamic faith was borne a short time later?

A lot of religions have similar moral codes. Some have one God, some several Gods. If the human brain was fully formed 50,000 years ago, it is most likely Gods, prophets, messengers, aliens, came to earth to teach the people. The three great libraries, probably at Alexandria, containing a history of the earth, were destroyed at the time of the Spanish inquisition, so we shall never know.

If we could find some common ground for religions to unite, we can consider the major problems of the world and ask religions to unite as one voice to help solve these problems. Most religions agree to the following *PRINCIPLES*:

1 God made us in his own image.
2 There is life after death.
3 We need a good moral code – evil behaviour reflects in the next life.
4 There is good and evil in the world – we must stamp out evil.
5 We need to live with nature in harmony.
6 No religion should allow its members to go to WAR.
7 We need to control world population.
8 We need to control world health and poverty
9 We need to control education.

Let us form a WORLD CHURCH of GOD. Every religion can join, keeping its own identity. Unite in a common cause in world peace

A NEW MORAL CODE

1 Be generous and helpful to your neighbour.
2 Do not kill another person.
3 Have a good code of behaviour in public life and in private and social life.
4 That your character can be respected to set a good example to others.
5 That you have a respect for your rulers and elders.
6 To strive to become a better person through learning and self examination.
7 To become a good family man.

Chapter 9

The future for planet earth

E arth is so vulnerable. Can we expect our beautiful planet to survive for us to enjoy it? What unseen danger might descend from the skies to bombard us or lie beneath the earth's surface to cover us? Many factors influence our survival whether it is from pollution to our atmosphere or damage to our soil and seas.

Let us examine the problems:

(1) Dangers from the sky. (2) Dangers from within earth's atmosphere.

1 Dangers from the sky must include:
(a) The Sun. (b) Comets. (c) Rocks from Space. (d) Supernova Explosion. (e) The Moon's Orbit. (f) Wandering Black Holes. (g) Attacks from Aliens.

2 Dangers from within the earth's atmosphere and pollution

a) Volcanoes and heating of the oceans. b) Damage to our atmosphere. (c) Alteration in world climate. (d) Damage to our rainforests and ecology. (e) Damage to our water supplies and the oceans (marine pollution). (f) Damage to world health. (g) Depletion of our natural fuel resources and re-cycling. (h) dangers from man and his wars.

1(a) THE SUN

The sun throws out excess energy in the form of solar flares. This radiation, travelling at the speed of light, 186,000 miles/second heads for planet earth. We are protected by the Magnetosphere, the Van Allen belts, as mentioned in a previous chapter. The ozone layer protects us from ultraviolet light rays. Our sun appears stable for a very long time to come, so we will be heated daily by its brilliance. There is a space rocket heading toward the sun at this time, packed with scientific instruments, so we may learn more about our sun. The earth's magnetic field may change in the next two hundred years. This will affect our protection layer, the Van Allen belts. The northern lights may become visible anywhere on earth.

(b) COMETS

New ones are discovered from time to time. Old ones come back after many years, e.g. Haley's Comet. They orbit our sun and travel back into deep space, sometimes never to return. Their mass and density is small in comparison to our sun and our giant planet Jupiter has been known to capture them in its gravity. It is unlikely they would collide with planet earth. Their orbits in our solar system can be predicted with accuracy.

(c) ROCKS FROM SPACE

A rock collided with earth 25,000 years ago in the Arizona desert. It left a crater 3,000 feet in diameter. This is possible at any time. In 1971 a large rock went past the earth close enough for it to be photographed, and later shown on television. It would have caused considerable damage if it had hit the earth. Where do these rocks come from? Of the nine planets orbiting our sun at regular distances apart there is a gap between the planets Mars and Jupiter. Instead of a planet, in its place there are some 35,000 rocks. Some are minor planets, the largest is about 10,000 feet in diameter. One rock collision with earth was probably responsible for killing the dinosaurs 65 million years ago, when a mountain of rock, six miles across, hit the Gulf of Mexico. The impact of this rock would have sent a shock wave and fireball encircling the earth of great heat, scorching everything, and polluting the atmosphere. There would have been a Tsunami wave of over 1000 feet travelling around the globe at 500 miles per hour. There would be re-bounding secondary waves, and the waters gone ashore would have retreated with such violence, everything was sucked away. Landslides and mud slides would have occurred. The polluted atmosphere would have blotted out the sun, causing the earth's temperature to fall; there would be no sunlight for many years, no photosynthesis, the dying of vegetation, starvation, pollution of water supplies and disease for the remaining animals.

These impacts are remote and hopefully may not happen again for a long time. Two-thirds of the earth's surface is sea, so the chance of any impact hitting a densely populated area is remote. It is the damage to the atmosphere that is a serious problem. Rocks from space are a serious danger to earth.

(d) SUPERNOVA EXPLOSIONS
Scientists can monitor massive stars and tell us what chance we have of surviving an explosion of a nearby star.

(e) THE MOON'S ORBIT
The moon is stable in its elliptical orbit around the earth, but it could be knocked off course by a mountain size rock from the asteroid belt, causing the moon to come close to the earth. We would have to attach rockets to the moon to take it further away, otherwise we would have astronomically large tides.

(f) A WANDERING BLACK HOLE
The first indication would be alterations in planetary orbits followed by rocks and comets colliding with the earth having been drawn off course by the gravitational attraction of the black holes. I think it would be time to get into our spacecraft and see if we could travel to a more suitable solar system, to find another earth.

(g) ALIEN ATTACK
In 1985 at the Geneva Summit, Ronald Reagan asked Mikhail Gorbachov' If earth is faced with an invasion by extraterrestrials could America expect help from Russia?' Gorbachov replied 'the United States and the Soviet Union would join forces to repel such an invasion, but I do not think it is likely as they seem friendly.' What strange things are reported in the Moscow newspapers.

2 Dangers from within earth's atmosphere

(a) VOLCANOES
Earth undergoes periods of heating, periods of cooling. The last ice age only finished, 12,000 years ago, when Europe, Asia, and North America were covered. If the sun is giving us a constant source of light and heat, which it has for millions of years, what controls the temperature on earth? The answer is volcanoes, the sea and the atmosphere.

Volcanoes are dotted all over the earth. Most volcanoes are situated at the boundaries of earth's 15 tectonic plates. The Pacific has large numbers of volcanoes. Indonesia has 125 volcanoes the most famous is Krakatoa. In 1999 Krakatoa started building a new summit. Hawaii has much volcanic activity and the ocean floor is rising in this area of the Pacific suggesting a basalt lava flow under the sea. Kilawayo in Hawaii is one of the most active volcanoes on earth. Basalt lava flows are very fluid and emanate from deep inside the earth 45 miles deep. Mount St Helen in North America is very large and dangerous and has erupted violently. Mount Etna in Italy is one of the biggest volcanoes on earth, erupting lava from 60 miles below the surface. Volcanoes damage the atmosphere. Sulphur dioxide mixes with water to make corrosive sulphuric acid. The contaminated atmosphere blocks out the sun causing a fall in earth's temperature. In 1883 Krakotoa's eruption caused a fall in earth's temperature of 1½ degrees. A 3-degree drop in temperature is enough to start a new ice age. Climatic changes can occur very quickly. The period 1600–1750 was very cold and the sea around England frozen in many places. The present period of global warming causing the ice caps to melt could have an unexpected effect. The release of a lot of fresh water into the gulf stream could alter the direction of the gulf stream and weaken it, resulting in Europe freezing up. There is a balance between global warming and global cooling. The ice age appears too cold, but who is to say today's climate is too hot. The question is whether we can foresee why it might get hotter and whether we can do something about it.

HEATING OF THE OCEANS

Temperature readings of the Pacific show an increase in sea water temperature often associated with an El Nino Year. This hotter water drifts across the Pacific Ocean to the Western American seaboard, noticed first by the locals who rely on fishing, and the fish go elsewhere. This hotter water heats the air in contact with it More evaporation takes place from the ocean, more rainfall occurs, more violent storms take place. The underwater volcanoes and thermal vents in the Pacific Ocean are responsible.

Earthquakes

Earthquakes form at the boundaries of earth's tectonic plates as plates move. Together with volcanoes erupting at these boundaries, other disastrous effects are activated such as mud slides, land slides, tsunami waves and floods. Buildings can be constructed to withstand some of the ground movements. Einstein predicted the earth's crust could suffer a massive frictional slip.

Pollution

(b) DAMAGE TO OUR ATMOSPHERE

Volcanoes we have mentioned. Other major problems are: -
1 Damage to the ozone layer.
2 Production of Greenhouse gases.
3 Air pollutants. Forest fires.

The ozone layer.

This is a form of oxygen in the stratosphere which stops damaging ultraviolet rays from the sun reaching the earth in large amounts. There is a large hole in the ozone layer over Antarctica. Chemicals used on earth in refrigeration and aerosol cans could be causing this damage. There has been a reduction in the use of Chlorofluorocarbons since 1980.

Greenhouse Gases.

Gases that pollute the atmosphere are accumulating in the atmosphere, like Carbon Dioxide. This prevents heat escaping from the surface of the earth, causing the temperature to rise, a contributory factor in global warming. Forest fires do immense damage to the atmosphere.

(c) ALTERATION IN WORLD CLIMATE

We are in a period of global warming. Some areas of drought will have rain; some areas of rain will dry up. This means we may have to move to more suitable areas to live. Animals migrate. Global warming is affecting plant and animal life, from polar bears not having enough time to get food supplies for their young, to penguins in the Antarctic, and butterflies migrating north earlier in Europe. More hot air means more violent storms, tornadoes in America, as hot and cold air try to mix; also an alteration in areas of high pressure, causing a change in surface wind direction and ocean currents. The sea level is rising at the rate of one foot in 60 years.

(d) DAMAGE TO OUR RAINFOREST AND ECOLOGY

Deforestation is the name given to chopping down forests of trees, to supply the wood market with timber for building. If nothing is planted in its place, the top soil may be washed away and habitat lost. In a tropical rain-forest are found the greatest variety of plants and small animals. Everything grows quickly. It is estimated 40% of our rain forests will be lost by 2050.

ECOLOGY

This is the study of how animals and plants fit into their environment. Organisms need food and materials to survive. This alters the food chain, and others have to adapt to the natural world. Deforestation deprives animals of this nature environment.

(e) Damage to our water supplies and the oceans (marine pollution)

Everyone needs clean water to drink. Disease is spread by contaminated water. When the atomic reactor at the nuclear power plant at Chernoval went out of control, the radiation damage was enormous, not only from the air, but from the ground. The seepage into the earth's water led to polluted water being spread over one hundred miles away. Hundreds of people have been contaminated and many have developed cancer and died. How long does radiation pollution last? We now make so much rubbish, the ocean's become the dumping grounds for everything, from local shore rubbish, ships at sea, to nuclear waste. Can we identify the types of pollution so we control them? Yes we can. Here is a short list: sewage, dredging dumping, mining, chemical industries, oil products, food processing, radio-active waste, water from pesticides and fertilizers and building materials, etc.

(f) DAMAGE TO WORLD HEALTH

With such a large increase in world population, so much poverty, so much pollution of both the atmosphere and water supplies, leading to pollution of oceans, is it surprising we have health problems of aids, cancer, heart disease, blood disorders, typhoid, mosquito carrying illnesses like denge fever?. Is denge not like malaria? There were 50 million cases two years ago, 600,000 cases in America and the Caribbean. Of the four strains, the haemorrhagic strain is most dangerous, which kills 30% of its victims. To make a vaccine they would have to make one to cover all four stains. It is now worldwide, carried by car and transport lorry tyres that have driven through water infected with mosquito aedes aegypti.

Chemical reactions produce the building blocks of life and like a jigsaw puzzle, if you remove one piece you have to repair the gap. This stable form becomes established over a period of time. When penicillin was invented in 1925, its widespread use led to the problems of resistance organisms. Bacteria learnt how to bypass the way that penicillin worked. One day bacteria will become so resistant, no drugs will kill it.

In the Caribbean I was talking to an elderly man. He told me, my mother taught me the use of twelve different substance from plant leaves, stems and roots to treat my illness. For example the sap from a cut stem of aloa has healing properties. Sometimes I think we are fighting nature, not joining nature.

(g) Depletion on our natural fuel resources

We could use up our supply of fossil fuels such as coal, oil and gas and grind to a halt. Other sources of energy must be found such as solar energy, wind power, tidal power, and controlled nuclear power.

RECYCLING RUBBISH

Glass, metal and paper all have good recycling potential. Plastics are difficult to recycle. Glass fibre does not deteriorate readily and it is illegal to burn it because of atmosphere pollution. What is going to happen to millions of glass fibre yachts? Are we going to have to sink them at sea when they are too old to use?

(h) Danger from man and his wars

At last the superpowers are dealing with acts of terrorism, extremists, fanatics, who wish to die for their beliefs, who wish to start wars. In this dangerous age when nuclear bombs can be made, and germ warfare has been used, the need to oversee trouble spots and send in negotiators, and inspectors, has given confidence that a third world war of nuclear proportions can be diverted. Now the cold war between Russia and America has finished, and China, Europe, Russia and America are sitting down to questions of peace keeping, lets have peace not wars.

Chapter 10

(a) Solving man's problems in developing the environment and adapting to it.

(b) The effect new discoveries will have on mankind in the next *50* years.

(a) POPULATION CONTROL

The large increase in world population in the last 50 years is expected to continue unless some controls are exercised. With better food, medical care, lower infant mortality rate, it is predicted the count in year 2000 of six billion will rise to nine billion by 2050 and twenty billion by the year 2300. Do we want to finish our days like a lot of rats in a cage with no fuels, no food, and no space left on earth, where psychologically we are at the bottom of the hill, fighting for existence, unable to climb back up?

China has taken drastic action to reduce their population of 1,250,000,000 in year 2000. No other country has had to cope with such a large population, except for India, with a population of 950,000,000. The USA is next largest with 265,000,000.

By the year 2025, the developed world will have a larger population of people over 65 to under 15. The underdeveloped countries, where poverty is the norm and there is a lack of money for food and medical care, e.g. Africa the Middle East, India, parts of South America and Asia, will continue to produce a larger child population. It is predicted, with birth control, the population of the developed world will level off in the next 10 years and start to go down. It will take a while because people are living longer. In Japan, Sweden, the UK and USA men can expect to live to 76, women 81. Everywhere in the world women outlive men by about 4 years or more. In the underdeveloped world birth rate is rising fast. The reasons are:-

1 No money for birth control pill.
2 Not enough money for medical treatment.
3 Lack of money for food.
4 Higher death rate, much higher birth rate.

5   Religious reasons preventing birth control.

Is it right a mother gives birth knowing her baby has a *50%* chance of dying under the age of 5 years? What right have we got to say you can't have a baby? The answer is take the birth control pill, keep the population down, then we can feed them, educate them, and provide some medical care. Surely religion can see the sense in birth control. In India food supplies sent in by charities have been seen stock piled in warehouses and not distributed to the people. Life expectancy in India is 55. Even if the world's population stopped at 9 billion, it would take another 20 years before it started to drop because people are living longer. An acceptable level to maintain with earth's resources is 5 billion.

EDUCATION

Having controlled the world population levels it is not such a big job to educate everyone to read and write. With global communication, the internet using universal language, English, and with world Television coverage by satellite, education can reach most people. There is some mental effort needed in learning and some discipline required. Examinations set goals and standards to be achieved. What subjects should be taught?

1   *A moral code, e.g.*
    (a) To be kind and helpful to other people
    (b) Do not kill another person; negotiate; don't fight.
    (c) To have a good code of behaviour and your character can be respected to set a good example to others.
    (d) To respect your rulers and elders.
    (e) To become a better person through learning and self examination.
    (f) To live in harmony with nature.
    (g) How to be a good family man.
2   How to stay healthy – how the body works – diet – exercise.
3   Music and the arts – learning to play the piano, violin, wind instrument is good for training the brain in co-ordination and laying down memory patterns.
4   Learning school subjects of languages, mathematics, sciences, arts, history, religion etc.
5   Learning a trade.

WORK SHARING

The need to have a job is very important. Long periods of unemployment are demoralising. It is better for the community that everyone plays a part, even if it means a short working week. Perhaps everyone should do one day a week helping the community, helping to run public services.

GLOBAL AGREEMENT ON ARMS CONTROL

We cannot allow nuclear weapons, germ warfare, weapons of mass destruction to get into the hands of terrorists. This requires the superpowers to unite in discussions and actions.

WORLD HEALTH

We are looking toward the superpowers to discuss the help needed for the poor and how to finance it. We need a shared knowledge on genetics, cures for illnesses, a control data bank on research, results and treatments.

BOUNDARY DISPUTES

We need a world peace keeping committee to settle boundaries between states. So much fighting starts across the border.

RELIGION

Use religion as a tool to unite and not divide. Most religions have common ground to unite in a common cause in keeping world peace. The problem is the fanatics, the fundamentalists. These groups need to be isolated within a community and educated.

CONTROLLING POLLUTION

We have to identify sources of pollution and areas most affected. Gases damaging our ozone layer are chlorofluorocarbons used in refrigerators, aerosol cans, air condition units etc. Other gases, chiefly carbon dioxide are contributing to the greenhouse effect. Other dangerous gases are sulphur dioxide and nitrogen oxides which form sulphuric acid and nitric acid. This rains down on us from the sky as ACID rain damaging the ground, plant and trees, which adds to deforestation with man's need to cut down trees for commercial purposes. This acid rain also pollutes water and the sea, killing fish. The gases responsible for air pollution are carbon dioxide and sulphur dioxide, methane and nitrogen oxides.

The principle areas which need control are:

a) Power stations and factories cause most of the problems.
b) Forest fires and domestic burning.
c) Motor vehicles, aircraft, ships.

The difficulty of controlling these areas are commercial interests.

POLLUTION OF WATER AND THE SEAS

Everything drains into the sea. The seas are becoming more pollutant. Prevailing winds and surface drift currents carry the pollutants. The Mediterranean is becoming a stagnant sewage. Major sources of pollution come from food processing and sewage disposal. Close second is metal and chemical industries, petrochemical and oil. Radioactive wastes are an unknown hazard. How long will containers hold the radiation when radioactive substances have a long half life of decay.

ADAPTING OURSELVES BETTER TO NATURE AND THE NATURAL ENVIRONMENT

This is tied up with population control, pollution control and ecology. Ecology is the study of habitat where plant and animals live in an environment and have to cope with the food chain and predators. Chemistry produced the start of life forms from sulphur products from volcanoes. The food chain is a continuing chemical reaction to extract the nutrients.

FINDING NEW SOURCES OF ENERGY

When all the coal, oil and gas products, including natural gas, are used up where will our energy come from? Hydroelectric power.

Hydroelectric power was first commercially used by Aristide Berges at his paper making factory in France in 1867. Electricity is made by the energy of flowing waters. In 1966 a hydroelectric dam was built on the river Rance at St Malo in Northern France, which makes use of the large tidal range. Water at high tide is trapped and drives turbines on the falling tide. We need to build large dams high in the mountains to catch the rain. The reservoir will provide water and hydroelectric power. Energy from wave motion in the sea has not so far produced electricity to be commercially viable. One mile off the coast of Lynmouth in North Devon a new experimental rig has been installed. A single pile driven into the seabed supports a large propeller. This is driven by the strong tidal steams in the Bristol Channel to produce electricity.

SOLAR ENERGY

The sun's rays are used to change light energy into electrical energy. They are used in satellites in orbit to run electrical equipment. They can charge batteries, so one day they may be able to run electric cars.

SOLAR FURNACE

The sun's rays are concentrated onto water by curved mirrors. This drives a steam turbine which makes electricity. The earth's heat from volcanoes and hot springs could be harnessed to produce steam and drive a steam turbine to make electricity.

NUCLEAR ENERGY

Nuclear power stations provide electricity. The energy released from nuclear reactions produces steam to drive a steam turbine which makes electricity. The first one was built in America in 1942. How great it would be if nuclear energy could be stored in batteries and the energy released slowly. Power stations can provide hot water for domestic use, e.g. Battersea power station in London.

WIND POWER

The Dutch first produced electricity from wind power. It was considered necessary to find a site where the wind strength averaged 26 knots to make it commercially viable. A windmill was set up on the Bloody Foreland in North West Ireland, Donegal. Since then wind farms have been developed with thousands of propellers which drive electric generators producing commercial electricity. Long distance sailing yachts have a simple wind generator of one propeller and solar panels to charge their batteries. The development of better stored electrical power in batteries is the future for pollution control from car exhausts.

Atmospheric pollution

People living in cities suffer from respiratory problems.

Chapter 10 (b)

The effect new discoveries will have on
mankind in the next fifty years

The unrest, terrorism, the threat of wars, will dominate the first few years from 2000, scientific attention will be given to early surveillance systems, war machines, blocking radar traps, missile defence and elimination of weapons of mass destruction in the hands of aggressive forces. By 2005, when world leaders have controlled terrorism and settled boundary disputes, the science of weapons of war may be put to good peaceful use. The capabilities of mass destruction of mankind in the wrong hands is a terrifying reality when atomic bombs and germ warfare can be delivered to their target by conventional aircraft, and aeroplanes do not need pilots to take off to-day.

The continuing research into better computers, genetics, DNA sequencing, quantum physics and bio engineering must continue. You cannot halt the big wheel of knowledge, the never ending research into new discoveries. We are going to find life in the universe.

At the top of the list is controlling world population. Then we can provide clean water supplies, food, education, health, better homes and living standards, teaching people to be independent and self sufficient. Next on our list is controlling pollution and ecology. We must stop rain forests being cut down, and our habitat de-nuded. At the current rate 40% of our rain forests will be lost by 2050, together with 40% of our species. The earth will dry up, leaving land not capable of supporting crops. We must understand our environment. No longer can forest fires be allowed to rage out of control, causing smog, poor air quality. Acid rain attacks everything and finally lands on the soil to pollute our crops or land in the sea to damage marine life. Aeroplanes will have better fuel and by 2050 will be flying faster at around 5,000 mph. Cars will be less pollutant, running on stored energy, batteries.

Where will all our energy come from? Oil and natural gas will eventually be in short supply. Reservoirs in the mountains will provide hydroelectric power, and making use of tidal energy to turn turbines e.g. St Malo in France, is another source of power. Windmill power is on the increase and solar panels. Can we not use hot earth vents for heating? Controlled nuclear power is still a source of vast amounts of power.

Cars are the world's biggest manufacturing industry. Is it a dream to think that one day we will have stored nuclear energy in batteries, to run our cars pollutant free? Too many cars and we will forever be in traffic queues. Would it not be better to develop inter city trains that travel at 400 mph? Also to keep cars out of city centres, park outside and take an electric train or bus/coach the last three miles to the city centre. There you can use taxis, buses, trains or electric bicycles.

The effects of tiredness and alcohol on driving still causes accidents and deaths. Mobile phones in cars will be banned. Cars can be made safer by monitoring the driver's mental health condition. Global positioning systems can be carried to give precise positioning to avoid getting lost. A track plotter can indicate the course taken on a map. A car could be guided along the road by computer automation using magnetic strips set in the road. This could help traffic flow in congested areas.

Computers

Computers are used in 1) industry, 2) in communication, 3) in travel, 4) in molecular biology, 5) in meteorology, 6) in physics and the properties of matter, 7) in medicine, 8) in the home. Computers are continuously changing in their design. Present computers will be limited by the physics of their design. Computers will become more powerful, smaller and cheaper. The future of computers is when computers are connected to each other. Microsoft has been ahead in the PC era. The development of the Macintosh Computer used ideas developed in the windows programs. Now the Internet is more widely used. Advances in computers has led to better bank cards, cash machines, the ability to do banking without using cash. This will eventually account for 80% of all transactions. The first computers were produced commercially in 1972.

The new scientific revolution started in 1945 with the exploding of an atomic bomb, the development of radar, radio and TV, x-rays, the transistor and the quantum theory. Typewriters were superseded by word processors and the CD.

We have come a long way since 1945 which makes certain predictions for the future to 2050 more likely. In 1945 the transistor was invented, which makes computers. The quantum theory explained what was happening. Energy is omitted in bursts and can be measured, 'Quanta'. Subatomic particles obey well defined laws, and can be predicted in their behaviour. Electricity is a movement of electrons, and transistors can amplify these tiny signals, and millions of tiny transistors form computers. This was possible due to the development of tiny circuits being imprinted on a silicone wafer using beams of light called photolithography. Together with lasers and x-ray crystallography, the properties of matter can be predicted. The structure of the DNA molecule and its double helical nature was discovered. This has led to the complete DNA coding of humans and many living organisms.

The Internet will be the world communication and knowledge, being fed by computers. It will become the world market place, because computing access will be available everywhere, and the ease of use, e.g. pointing a 'mouse' at the pictograms to open programs and use them, is readily understandable. Will we be able to talk to the Internet in the future, ask it a question and have an electronic robot answer in English, or will we rely on E-mail?

The Internet came into commercial operation in 1988. By 2020 90% of world knowledge will be accessible to nearly everyone on earth. The ability to sell goods, market holidays at a cheaper price will help to keep prices down, and make the Internet attractive from a business point of view. It can advertise jobs available, and stimulate world economy.

Many people have TV. There will be a merger between the Internet and TV. TVs will change. Instead of producing a picture in a cathode ray tube which scans 525 lines across the picture screen, it will be 'digital'. The resolution will be much better and transmission not affected by distance. We will have to buy a new TV at sometime in the future, about 2006, and have an adapter to connect it with the Internet. By then a lot of people will have access to the Internet in their homes. There is a lot of junk on the Internet, and unless this can be eradicated, the appeal of the Internet is limited. It may be possible to train a robot to do our searches and cut out the junk. To make artificial intelligence that can think for itself is a long way off. The computer is not restricted in its input as is the human brain. By connecting computers together, it may be that one computer can talk to another computer. Will we eventually make a computer that can read our thoughts? Will these computers be able to access our brains, psychokinesis?

World TV can be relayed through satellites, at a price. Cable wires are

160

convenient, and cable companies are offering fast Internet access as well as TV, again at a price. Transmitting from satellites in space has important implications. The global positioning system (GPS) works out your position in latitude and longitude to the nearest 30 feet. Distress from ships at sea, called global maritime distress and safety systems, operate through satellites to a central surveillance control room. Ships identity is coded and distress activates a response for rescue to a known position. Laser fibre optics may be the means of transmitting for the future and be part of the Internet make up. Fibre Optics will replace copper wires.

Robots Automations and Artificial Intelligence
Industrial robots are like big toys. They obey instructions written on computer discs. They are used in atomic power stations, in space probes to planets. They can be made more sophisticated to work in commercial enterprises, mobile robots which move and miss obstacles and made to cut grass.

The Human Brain – The Next 50 Years
Of the 100,000 genes that make us, it is estimated the brain contains 3,195 genes. How do all these genes react with each other? We don't know. There are millions of nerve fibres coming up the spinal cord into the brain. Do we know what they are all doing? We do not know. We do know the function of some areas of the brain, and some of the neurotransmitters, the chemicals controlling brain function.

Most parts of the brain have sensors representing sensory ending from the body. The fingers have pain, temperature, touch, visual, and pressure points, all represented in the brain and in the memory banks. The brain surface does not have sensory endings. A small stimulation produces a small response in the brain. A large stimulation can cause the whole brain cortex to respond. When you go to sleep the energy not analysed and dissipated will be released so that neuron route can be used again. This may produce dreams, a bizarre muddle of events, as the brain sorts out the return to normal neurone transmission.

The brain also has unwanted characteristics from millions of years of development, genes that may be dormant or changed in function, e.g. today we are not being hunted; we do not need the sudden release of adrenaline, muscular tone to run away from danger Can we ever put consciousness into a robot? Can we expect a robot to have a level of awareness, with the ability to recall from memory and understand the response to give; a robot that thinks! We need to build a human brain to understand it. First we must understand the small brain of an insect and build upon this knowledge toward a larger brain.

Do flowers think or have a level of consciousness? Are they aware of their surroundings? Flowers do not have neurons and a brain centre, so their reactions are to the environment by sensors directly acting on a response centre which may close petals, leaves.

If the brain has 3,195 genes, the liver 2,091 genes etc, how do we build a real one? We do not know. If I were to build a house, I would need a master plan. Then various specialist would build each section, carpenters, plumbers, bricklayers, electricians, roofing contractors etc. Do we have a master gene, and chemical and electrical genes? If our eye first developed from a fish's eye 500 million years ago we could say we have traced our master gene. We know a lot about sight so perhaps we could make a lens and a retina, provided the connections were still made to the visual cortex. In the next 50 years we can see advances in understanding how the

Genes pass on the inherited characteristics, even to the colour of our eyes. They also pass on the inherited diseases, e.g. cystic fibrosis. Now we have a map of the body showing where all 100,000 genes are located, we will be able to have our own genetic coding printed on a CD and analyse it on a computer. Your doctor can tell you if you have genetic diseases and preventive treatments necessary.

DNA. This consists of two coiled strands in the form of a double helix. The genes lie along these strands, and make up 23 pairs of chromosomes locked inside the cell nucleus. Nucleic acids form the DNA and the precise sequencing of these nucleic acids, ATCG along the DNA are called Schrodinger Genetic Code. Some *50* amino acids create the DNA, One example RNA – Ribonucleic Acid. Scientists must investigate how genes work in our bodies, and the part played by 100,000 protein molecules that make us.

ADVANCES IN MEDICINE IN THE NEXT 50 YEARS

Diagnosis

1   Blood analysis and body scans will be supplemented by a microchip that can screen your DNA and test for selected genes, test for inherited diseases, HIV and cancer. DNA examination of an individual for crime is better than fingerprinting. There may be home test kits linked by computer to the doctor's surgery.

162

2 Better medicines and understanding dietary needs at different ages – educating the public.

3 Improved intakes – water purification and research into the foods we eat – Carcinogenic agents?

4 Replacement organs.

5 Improving immunity – why do crocodiles have total immunity?

6 Prolonging life – how do we make the tissues stay young?

7 Understanding how genes perform in our bodies, using research in animals and yeasts.

8 Cell reproduction – cancer and its treatment. Mutations in P53 and the way it causes cancer, e.g. Benzoa Pyrene Doil Epoxide found in tobacco smoke causes lung cancer.

9 How to find cancer cells – early warning signs – it may be possible to make a radio active isotope of the enzyme Telumerase, only found in cancer cells, which caps the end of chromosome to prevent it dying, thereby giving long life to a cancer cell at the expense of the other cells.

10 The elimination of certain diseases, e.g. small pox, poliomyelitis, leprosy etc, some diseases have crossed over from animals to humans, e.g. the plague transmitted by the fleas in rats.

11 The battle to find the weak spot in virus micro-biology to kill resistant strains.

12 When our bodies move, the static electric charge around our bodies from electrons moves and produces an aura. How this aura can help in diagnosis.

Ageing. Can we Live Longer in the Next 50 Years?

The four main lines of research are centred around:-

1 Looking for ageing genes.

2 Replacement organs, transplants, rejections.

3 Substances shown to prolong life.

4 Diet – the body requirements – fat genes – the control of our body weight will increase life expectancy.

Genes Producing Ageing

Experiments on mice indicate there are ageing genes. This must not be confused with the results of experiments on low calorie intake, which show mice live much longer at the expense of a lowered basal metabolic rate. This lowered activity leads to an inability to reproduce.

Replacement Organs
There are two ways this can go ahead:-
1 Organs built in the laboratory.
2 To stimulate a body to grow new organs. The information to grow new organs is locked in the DNA of specialist cells. Can we keep a data bank of specialist cells and grow them in animals?

Substances Shown to Prolong Life
This is a controversial subject. The Russians have indicated Pangamic Acid = B15 is the answer. Other findings suggest antioxidants, e.g. vitamin E helps. No doubt there are arguments for many substances, which can be considered under the heading dietary needs.

Diet and Bodily Needs
There is evidence less food intake and a quieter life leads to a longer life. Can we slow down our basal metabolic rate? There are some definite indicators relating to certain organs and the ageing process. After the age of *50* the pancreas may not be producing enough insulin to cope with the sugar intake. Cut back on sweets and fatty products – less fried food and no sugar in tea or coffee. Why not cut out the tea and coffee and have fresh orange juice. Your immediate energy needs can come from other foods than sugar. Eat a little, but more often.

Some foods are good for you, some are bad. Some people have allergies and special dietary requirements. Lets start with a list of vitamins, what they do, and what foods to find them.

Vitamin A	Found in butter, eggs, liver, animal fats. Deficiency leads to night blindness as vitamin A is necessary for the regeneration of visual purple.
Vitamin B	Found in vegetable foods, cereals, bread, nuts, seeds, beans, liver, meat, fish etc. Vitamin B can be synthesised by bacteria living in the small intestine. It is unlikely to find a deficiency unless prolonged use of antibiotics have sterilized the alimentary canal.
Vitamin Bi	Thiamin – cell metabolism.
Vitamin B2	Riboflavin – cell respiration – necessary for energy – lack

164

	of causes cracked lips, depression, tiredness and skin complaints.
Vitamin B3	Niacin – necessary for energy. Deficiency leads to depression and muscular weakness, gastro intestinal disorders.
Vitamin B5	Pantothenic Acid – helps in immune systems.
Vitamin B6	Pyridoxine – helps in healthy skin and in neuritis.
Vitamin B12	Cyanocobalamin – helps in pernicious anaemia.
Vitamin C	Ascorbic Acid – found in fresh fruit and vegetable, tomatoes, potatoes. Helps to form collagen and reticular fibres. Also involved in the immune system. Drinkers need extra Vitamin C. Lack of produces scurvy and poor wound healing.
Vitamin D	It is synthesised in the skin by the action of ultra-violet light on a 'pro-vitamin'. Prevents osteoporosis and helps in the immune system. A lack of it in infancy leads to rickets because there is a failure to absorb and utilize calcium and phosphorus.
Vitamin E	Tocopherol – helps to stop ageing – found in nuts, vegetable oils, green leafy vegetables.
Vitamin K	Found in spinach and other vegetables. It is essential for the formation of prothrombin in the liver. Aids blood clotting – also in bananas.
Vitamin B15	Pangamic acid – Russian's claim it oxygenates the tissues and keeps them young.
Vitamin B17	Amydallin – found naturally in kernels of apricots – tastes bitter – taken as a supplement in cancer treatment – theory – the cyanide radical is liberated in the cancer cell by oxidase and kills the cancer cell, being excreted harmlessly. Birds don't get cancer.

Minerals

Again another big subject. To make the amino acids, the neurotransmitters in the brain, bones, and everything else, we cannot afford to be seriously deficient in any one and above all our immune system, hormones, enzymes, need to be made. Sodium and potassium make nerve cells.

In Summary

A balanced diet, low in salt and sugar, with plenty of fresh fruit and vegetables can be recommended. Weigh yourself and keep your weight down; remember this is the secret to longer life. Supplements of vitamins and minerals can be good, but too many are bad for you. Alcohol is not good for you, so keep it to a minimum. It is suggested red wine may be least harmful.

By 2050 we should be able to live to be 100, with some people reaching 130. Commercial interests come into most modern living, and you can probably live longer if you can afford it. Commercial interests come into the foods we eat. Are genetically grown foods going to be any better for us if the soil they are grown on is contaminated? What preservatives are used to keep them fresh once they have reached the market place? What is the long term effects of pesticides used on crops? By the year 2050 we should know some of the answers.

With gene transfers we can by pass millions of years of evolution and create new plants and new animals on earth. We have been created, and now we are going to be the creators. What shall we make? a cuddly baby dinosaur? If we could, the new transient genetic code would be passed on permanently in the new species. New plant strains are being produced and new seeds, and new food crops, with disease resistant qualities. In 1997 a sheep was cloned, called Dolly, the first time it had ever been done, to produce a carbon copy. Let us hope it does not happen to humans. Genetic engineering can change the human genome and change the human race STOP! It is time to put the brakes on?

We need to develop high quality relationships, a world where we can all interact together, to do things together, to want to spend more time together. By 2020 this will be happening.

New Energy in the Next 50 Years?
The future for travel and spacecraft may lie in the following. Can we overcome friction? This is a waste of energy in mechanical systems. The answer lies in the ability to hover, using magnetic fields created by super conducting supermagnets. Super conductivity will overcome the heat generated by electrical resistance. Superconductors must be made that can operate at room temperature and not in very cold liquids. This is a major line of research today.

Another science is developing called Nanotechnology. Scientists are trying to make machines from atoms that work by moving parts by arranging the atoms into gears. Both these new lines of research can lead to enormous implications for the future. 1) that we can overcome friction, 2) that we can use atoms to make machines with enormous power, 3) that we can 'pass by' the problems of heat generated in electrical resistance; even computers heat up.

There are problems to be solved, 1) can superconductors be made at room temperature? 2) can superconductors retain their superconductivity properties in the presence of large magnetic fields? Observers of UFO's have reported the whining noise of electric generators starting up before super magnetic fields hover the craft. Are they using these principals for take off? They are producing a lot of laser light, and I expect their command vessels are using the energy of a star, nuclear fusion.

By 2050 it may be possible to operate commercial fusion plants to light our cities. If this technology has not arrived by them, we can expect to see widespread use of solar panels to produce electricity to individual homes, as solar panels become more efficient. If battery power and solar panels become more efficient, we might see the return of electric cars, covered in solar panels, useful for short journey, where acceleration is not important. Instead of buying petrol at the gas station, you would buy energy stored cells to plug in, swapping your used ones for re-charging.

INTO SPACE

History since 1947.
To leave the earth's gravitational pull and go into orbit around the earth, was occupying the scientists at this time. The speeds required by rockets were known, and German scientists after the second world war were ahead in rocketry. The re-entry problems required materials that could burn off the heat, as craft bounced back into the earth's atmosphere, causing great heat friction. The mathematics were in place, the rocket technology followed. On 4 October 1957 the Soviet Union launched the first artificial satellite into orbit. It functioned 21 days. The space race was on. The cold war between America and Russia was aggravated by the possibility such satellites could carry atomic bombs, which accelerated the American space program. Weightlessness posed a problem, and the Russians launched a dog into orbit to study the problem. The Russians followed this by Vostox I, launched on 12 April 1961 and the first man in space Yuri Gagarin, returned safely after one orbit of the earth.

The Americans also launched satellites and in January 1961 they launched a chimpanzee into orbit to study weightlessness. On 20 February 1962, John Glenn, became the first American to orbit the earth. The next

'Space Race' was to land a man on the moon, and to see the far side of the moon we had not seen. In July 1969 Apollo II landed two men on the moon and incredible pictures of Blue Planet Earth were taken from the moon.

The plans for a space station were formed, a place where scientists could dock, and work, living for longer periods of time in space. The Russians and Americans joined forces in space. The first space walks took place and American 'Buzz' Aldrin spent 5½ hours outside his spacecraft. These space walks were necessary prior to the launch of Apollo 11 on 16 July 1969 to the moon. It took 4 days to reach the moon, and Neil Armstrong and Buzz Aldrin stepped on the moon.

Space Stations

Salyut was the Soviet Union's first space station in April 1971. Seven stations were set up in the course of 15 years. Following on, the Russians launched the Mir Space Station, adding modules over many years, the last module in 1996. Skylab was the American equivalent, launched in May 1973, eventually destroyed in 1979 when it re-entered the earth's atmosphere.

The International Space Station

This is a joint effort of many countries. It started being assembled in November 1998 and will eventually be enormous, 450 tons, when finished. There will be 7 research laboratories and living space, orbiting 200 miles above the earth. The American's designed a re-usable space shuttle in the early 1970's, which came into operation in 1981, and returned to earth to land in a conventional way for re-use. It has flown many times. It suffers from the danger of being damaged by floating debris in space.

The Hubble Space Telescope was originally faulty but thanks to the knowledge gained by space-walks the telescope has been repaired and is working well. A new telescope 2½ times bigger than the Hubble telescope is scheduled to go into orbit one million miles above the earth's surface in 2007.

There is a lot of debris floating about in orbit above the earth. Although a lot of it is in small fragments it is presenting a danger on account of its speed at 18,000 miles per hour. The New International Space Station could be damaged by this debris or by small meteorites. Will the new space station become an expensive 'white elephant'? It is only scheduled to last 15 years. The US is designing a new type of space shuttle to supply it. New types of telescopes will be used on it. A manned space flight to mars may be launched from it. Mars may have contained life in the past. Ten space probes are investigating mars between 1997 and 2007. Also life may have existed on Europa, one of Jupiter's moons. Beagle two is a British robot launched on a Russian rocket to land on Mars, to probe the surface. Travelling at an average speed of 20,000 miles per hour it will take 18 months to reach Mars, which is 250 million miles away. It should arrive in 2004. A new satellite, swift satellite, is due to be launched in December 2003 to study gamma ray bursts.

New Propulsion Systems

A new engine is under test. It is called an Ion Drive. Atoms of xenon are electrically charged and driven out of the exhaust by a powerful magnetic field. Inter stellar travel will need to make use of nuclear fusion to utilize hydrogen, but until the problems of nuclear fusion can be solved, we must be concentrating on the ion engine. Solar cells provide electricity to heat up and ionize xenon, to be ejected at high velocity to provide the thrust, but only a moderate amount of thrust will be achieved. This means a combination of rockets must be used.

Earth-Like Planets in Space?

Yes, by the million, if the laws of 'average chance' prevail. We have a solar system like ours only 8.1 light years away orbiting the star Lalande. It has several planets. Perhaps one day we can visit it. Do we need to? One day we will need to, when earth is devastated by impact with a mountain of rock from the asteroid belt. Imagine it is five thousand years on. We have colonized a new planet called Bonito. The history books have been written, telling our civilisation how they came to be. A great spacecraft came from the sky and cultivated and built our planet, like their original planet they called earth, which lost its atmosphere in pollution.

I wonder if the great libraries of planet earth in Alexandria which were destroyed at the time of the Spanish Inquisition by the Roman Catholic Church, which contained the history of earth from the time of the beginning, would tell the same story!

Chapter 11

Observation and conclusion

In chapters one and two we have seen a vast universe, beyond comprehension, where the building blocks of life are made in stars. Planets orbit their stars at a certain distance and when the temperature is right oxygen and hydrogen make water, to form the oceans. Life is in abundance in the universe. Some life forms will be more advanced than us, some less, and some forms could be different. We may only just be at the bottom of the ladder of knowledge, a long way to climb up to the top, but in 2003 we have a bigger accumulated background of scientific knowledge to work from.

There is evidence we have been visited by aliens for thousands of years. In the last 55 years they have landed, keen to collect data, soil and human samples. They don't seem to want to communicate or give us knowledge. If they wanted to find out more about us they would talk to us. I think we are part of their experiment or 'whatever it is', they do not want us to know. If they are collecting soil samples this suggests they live on a platform where there is no soil. They have discovered how to travel the vast distances across space which is going to hinder us in space exploration in years ahead.

With limited range of vision, hearing and senses we are not processing enough data entering our brain to understand what is going on. The use of a computer can help, but can a computer think like a human brain? We need to build a human brain and body to go with it, to provide the sensory input. This would be too difficult to do.

The development of the music centre of the brain may be more important than has previously been recognised. Perhaps it does indirectly play a part in survival. Birds have perfect pitch so they can call a mate. Sound waves are an important part of communication. I would guess our music centre would have developed from singing.

The number of brain cells allocated to music is very small. This does not mean we cannot develop the area, because other areas become involved.

Music can be soothing, reassuring, rigorous, to bring us into action. Research at Harvard University has shown some classical music sounds to be soothing and assist healing processes. Pregnant mothers have put earphones on their bellies to play soothing music to their unborn. Why can our emotions be so deeply stirred by a concert of beautiful music? Children should be encouraged to take music as a subject at school and learn to play and appreciate music. Sound waves can be low pitched or high pitched and soft or loud. If soothing music is 'settling', 'comfortable', 'relaxing', to the brain, so more vigorous music could be associated with movement and bodily action. We can associate time signatures with physical feeling e.g. March time or Waltz time. By occupying your mind with movement and bodily action, music can distract your attention away from poverty and a harsh environment to help achieve happiness – dance the Samba – Latin American.

The church recognises evil forces, spirits that can possess people's brains, and the need for exorcism to drive them away. The Voodoo Cult or religion practices the art of communication with these spirits. Regression under hypnosis, when people are taken back to their childhood and asked what they were in a previous life can produce extraordinary tales which can be checked against history. Also associated with this is the ability to talk in many tongues. All that is happening is you are inviting a spirit from the past to speak through the person and tell of the circumstances at the time – Psychokinesis. We have the free will to create and choose. We can use the good forces from 'beyond' to heal through psychokinesis, e.g. faith healing. The church has said there are too many forces of evil in the world today, which need exorcising. If good and evil spirits can operate through our brains, this must lead to the possibility that an alien connection could reach us through our brains.

In nature I have tried to show action and reaction leads to a stable form illustrated so well in the decay of radioactive substances. Interfering with the stable form of nature will only produce a counter reaction. I illustrated my point with the natural attraction of certain chemicals in producing the building blocks of life. I do not believe the beauty and diversity of creation came by chance from disorder. I believe it came by design. I do not believe in an all powerful and important good god with so much evil and suffering in this world. Everything points to an experimental laboratory with things going wrong. I do believe we have been visited by aliens for thousands of years, and we have been taught by prophets from the past, telling us there is life after death.

171

What About the Future?

Did our creator pull the plug out on the dinosaurs 65 million years ago? Was the experiment completed? Alternatively; was our creator unable to stop the mountain of rock some 6 miles across from falling to earth in the Gulf of Mexico? Does this mean all our treasures of art, history and beautiful things are for nothing?

Over a life span of 100 years the chances of earth getting a bombardment of rock from the asteroid belt of a serious nature is only slight. Over 50,000 years, catastrophes are more likely. Hopefully our creator is not a restauranteur. Many of the things we cultivate we eat. I can understand what food is good for me and what is bad for my stomach. I cannot understand what is good for my soul and what is bad. If my soul comes from my brain, then music gives me pleasure, and horror movies give me nightmares. Am I to be judged in an after-life situation by my thoughts? Some religions offer a good code of practice, a set of standards to maintain in everyday living. Our creator cannot expect us to understand 'anything' too complicated. If religion is simple it can be followed by the poor, and comfort the sick.

I believe many secrets will be revealed when we understand more fully the properties of light and electromagnetic forces, and time dilation. It would be difficult to deny the possibility of life after death. Ghosts and poltergeists are appearing from time to time. I have watched a man who can see and communicate with these apparitions, give them a good 'telling off' and send them on their way, called exorcism. With so many unexplained things in our universe we must continue our researches and experiments. One day in 1000 million years time, when our sun expands as a red giant, we will need to get off the planet and find a better place to live – another greenhouse? In the meantime can I give my thoughts?

If a child asks an awkward question – why do we die and where do we go? I tell a story. God made a beautiful place called the Garden of Eden full of fruits, and made Adam and Eve to live in the garden. Adam and Eve had babies, and the babies grew up and had more babies, and soon there was not going to be enough room on earth for everyone to live forever. So God made a Chinese restaurant and put it in the Garden of Eden, and the food was so good everyone ate the food. Then God gave a new menu to the Chinamen and this caused everyone to die at 100 year's old, and go into the next life where there is more room.

1   Enjoy the beauty and diversity of creation.

2 Accept we are in an experimental greenhouse, we are part of the experiment.

3 Time on earth is short, i.e. in one year in a spacecraft could see the passing of 1000 years on earth, making the experiments from space very observable.

4 If we believe the bible and God made us in his own image, consider function as the chemistry of reaction producing an illusion of reality, and the appreciation of music and art as a higher function. This higher function is not associated with the aggressive needs of survival.

5 Now that man can survive without this aggressive force of survival, it is no longer required. Religion tells us of common code of practice in all religion of peace and love to one another. This is the present part of the greenhouse experiment. We are being conditioned by a code of practice, to love and be peaceful, or in scientific terms to genetically modify us to drive out our aggressive tendencies which are necessary for all developing things initially to survive. This accounts for the way the human brain has been formed in a complex way to take into account the appreciation of music, art and many things not associated with survival. The development of this non-aggressive function is paramount in the future, and the code of practice 'To Love Thy Neighbour' in religion is the ideal way to suppress this unwanted characteristic.

Let us put everything and everybody onto a balance and see what happens.

On the left side we will put:-
Pain, suffering, earthquakes, volcanoes, terrible storms, cold, ice and snow. Disease, hunger, thirst, uncertainty, death.

On the right side we will put:-
Love, happiness, fun, achievement, beauty, pleasure, reproduction and art.
Planet earth is in balance between good and evil.

Now take away everybody.

EVIL   GOOD
_____
                      ∧

The balance is the same.
Now add to the right side our addition to the beauty of creation, our music, art, poetry and achievements.
  The balance goes down on the right side.

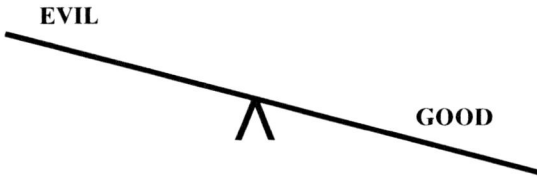

EVIL

                    GOOD
          ∧

The creator wants to share the beauty and diversity of creation with us.

Where is heaven?
In a place where there are no earthquakes, no volcanoes, no cold and suffering, no uncertainty, only stability.

Where is stability?
A vast artificial platform on which a vast city stands at the boundary layer of spacuum between gravity and anti-gravity.

Can we see it?
Yes, you just need a bigger telescope. Look for a small dense galaxy with no peripheral stars, the boundary layer between gravity and anti-gravity. There is no darkness. The light from other dense stars shines by day and night.

We have found the Perfect balance

$E = MC^2$ ——————————————————— SPACUUM ENERGY

If you go to heaven before me can you ask a question? Who made the big bang 16 billion years ago?

The answer

We are working on that problem and have made some good progress. In heaven we have a lot of very ancient documents. You are from planet earth. Perhaps you can help us decipher some of these documents. Is there more than one heaven, and are they all different?

ISBN 1412016640-1